# WIND**POWER**

Christopher Gillis

Schiffer® Publishing Ltd

4880 Lower Valley Road Atglen, Pennsylvania 19310

Copyright © 2008 by Christopher Gillis

Library of Congress Control Number: 2008924840

All rights reserved. No part of this work may be reproduced or used in any form or by any means—graphic, electronic, or mechanical, including photocopying or information storage and retrieval systems—without written permission from the publisher.

The scanning, uploading and distribution of this book or any part thereof via the Internet or via any other means without the permission of the publisher is illegal and punishable by law. Please purchase only authorized editions and do not participate in or encourage the electronic piracy of copyrighted materials.

"Schiffer," "Schiffer Publishing Ltd. & Design," and the "Design of pen and ink well" are registered trademarks of Schiffer Publishing Ltd.

Type set in Zurich BT

ISBN:978-0-7643-2969-2

Printed in China

Schiffer Books are available at special discounts for bulk purchases for sales promotions or premiums. Special editions, including personalized covers, corporate imprints, and excerpts can be created in large quantities for special needs. For more information contact the publisher:

Published by Schiffer Publishing Ltd.
4880 Lower Valley Road
Atglen, PA 19310
Phone: (610) 593-1777; Fax: (610) 593-2002
E-mail: Info@schifferbooks.com

For the largest selection of fine reference books on this and related subjects, please visit our web site at **www.schifferbooks.com**
We are always looking for people to write books on new and related subjects. If you have an idea for a book please contact us at the above address.

This book may be purchased from the publisher.
Include $5.00 for shipping.
Please try your bookstore first.
You may write for a free catalog.

In Europe, Schiffer books are distributed by
Bushwood Books
6 Marksbury Ave.
Kew Gardens
Surrey TW9 4JF England
Phone: 44 (0) 20 8392-8585; Fax: 44 (0) 20 8392-9876
E-mail: info@bushwoodbooks.co.uk
Website: www.bushwoodbooks.co.uk
Free postage in the U.K., Europe; air mail at cost.

# Contents

# Foreword

The wind power industry in America has had four distinct periods of development beginning with the early settlers bringing the European technology to America. The settlers found that the European designs did not provide the flexibility to capture and withstand the fickle weather, especially in the Midwest. The American windmills that were self-regulating and had pivoting blades for high winds were introduced in 1857. The Eclipse windmill was the first to use a solid wheel and a side vane to turn the rotor out of the wind as the rotor speed increased. These water pumping windmills provided year-round water supplies that allowed much of the Great Plains and Midwest to be settled between 1870 and 1920. The third distinct period was the use of wind power to generate electricity. Small units were installed to charge batteries that would supply lights for the kitchen and bathroom, with enough extra power to listen to the radio for an hour or so in the evening. The Jacobs Wind Electric Company reported selling tens of thousands of these units between 1931 and 1957. Most of these machines were removed when the Rural Electrification Administration (REA) installed electric power lines. With the increased cost of fuel in the 1970s, wind enthusiasts and the U.S. government started developing wind systems using modern aerodynamics learned from the aerospace industry. The first major installations occurred in California during the 1984 to 1986 time period when almost 15,000 turbines were installed.

The original wind research and development program of the U. S. Department of Energy contained six program missions. The wind characteristics and wind maps were studied by Battelle Northwest Laboratories. The Solar Energy Research Institute was responsible for innovative and unproven technologies. NASA designed and built four different prototype large machines ranging from 100 kilowatts to 2000 kilowatts. Sandia National Laboratories designed and developed the vertical-axis technology, leading to a commercial company installing almost 1,500 units in California during the 1980s. Rockwell International designed and developed small machines of 8, 15, and 40 kilowatts. Two partnering companies developed their designs into commercial products that were part of the California installations of the 1980s. Finally, the U.S. Department of Agriculture's Agricultural Research Service developed controller interfaces to allow these modern turbines to be used in rural and remote applications such as pumping water, cooling milk and farm products, and ventilating animal buildings. The federal wind research was consolidated when the National Renewable Energy Laboratory was established. Sandia National Laboratories and USDA's Agricultural Research Service continue as contributors to the federal wind research effort.

At the same time that wind research and development was started in the United States, the Europeans were also hard at work. When the California market opened in 1983, the Europeans were ready with their machines as well. About half of all machines installed in California were manufactured in Europe. Since the mid-1970s, the wind industry has been an international industry with international competition. The U.S. wind industry struggled through almost two decades of on-again and off-again energy policies, resulting in the Europeans taking the lead in wind turbine technology and at one time, leaving the United States with only one manufacturer of utility-sized machines. Several U.S. small machine manufacturers managed to produce enough revenue to remain in business through the 1990s. As world energy prices began to climb at the start of the 21st century, wind turbine manufacturers began to see increased sales as individual states started to establish renewable energy portfolio standards. These state renewable energy standards required electric utilities to add renewable energy to their generation mix. The wind industry was ready to meet this demand. Wind electric power production has increased at about 25 percent per year since 2002, making it the fastest growing power generation technology. Today, the leading manufacturers of utility wind turbines are located in Denmark, Germany, Spain, the United States, Japan, India, and the Netherlands. In 2007, the common size of turbines installed for utilities range from 1.5 to 2.5 megawatts. The machines have rotors over 300 feet in diameter and are on towers 280 feet tall. One megawatt will supply power for 350 to 400 average U.S. homes.

As we see the utilities endorse wind power, and wind farms or wind plants spring up all over the United States, small manufacturers are making machines for individual homes, schools, and small businesses. These units are used to offset high electrical costs and provide the individual owners with an opportunity to reduce emissions from coal-fired generating plants. Wind energy enthusiasts, whether owner, manufacturer, or supporter, have the feeling that they are making our world a better place to live and will help to protect our environment for years to come.

R. Nolan Clark
Agricultural Engineer
Conservation and Production Research Laboratory
U.S. Department of Agriculture
Agricultural Research Service
Bushland, Texas
September 2007

# Acknowledgements

The purpose of this book is to provide a mere overview of how wind energy and its technology have evolved in application, not just in North America but also globally through 2007. It's neither a definitive historical account nor a technical manual. The goal is simply to provide a pathway for readers to learn the basics about this increasingly important energy source and generate sufficient interest to seek out additional works by noted authors of the field and Internet-based sources for more in-depth knowledge and understanding.

It must also be noted that this book would not have been possible without the kind assistance and support of numerous government and industry experts, historians, and friends.

In the area of wind energy history and early pioneers, I received support from T. Lindsay Baker, author and editor of the *Windmillers' Gazette*; Frans Brouwers, editor of *Levende Molens*; Povl-Otto Nissen of the Poul la Cour Foundation; Coy Harris, executive director of the American Wind Power Center; Bob Bracher, president, and Gene Kincaid, Webmaster, for Aermotor Windmill Co.; Norma Sue Hanson, head of special collections at Case Western Reserve University's Kelvin Smith Library; Bruce Yelovich, research librarian at Mount St. Mary's University; Dietmar Jost of Düsseldorf, Germany; Peter Neyens and Johan Copermans of Belgium; and the books of American historians Paul Gipe and Robert Righter.

While researching the history of small wind energy systems, I encountered Craig Toepfer, who unselfishly shared with me his deep historical knowledge, his experiences working with Marcellus Jacobs, founder of the former Jacobs Wind Electric Co., and photographs. I am extremely grateful to him. I would also express my thanks to Michael Bergey, president of Bergey Windpower; and Andy Kruse, co-founder of Southwest Windpower, for sharing their time and knowledge about the modern small wind energy sector.

From the government, I would like to thank R. Nolan Clark, agricultural engineer for the U.S. Department of Agriculture's Conservation and Production Research Laboratory, who wrote the Foreword for this book; Marvin Smith and Deborah Demaline from the NASA Glenn Research Center; Walter Musial and Kathleen O'Dell of the U.S. National Renewable Energy Laboratory; and Chris Burroughs of Sandia National Laboratories.

My ability to write knowledgably about the transportation of wind turbine components both landside and offshore comes from my 13 years of writing for *American Shipper* magazine. Through this venue, I encountered experts in this field from firms such as LoneStar Transportation, American Transport Systems, A2SEA, Intermarine and MPI Offshore Ltd. Turbine and related component manufacturers Clipper Windpower, Suzlon, Vestas, Horizon Wind Energy, and Knight & Carver were generous with both information and photographs of their products. I would like to offer a special thanks to Brad Adams, president of Whitewater Wind Energy, who walked me through the process of turbine construction, operations and maintenance.

For the most up-to-date facts, figures, and photos related to the wind energy industry, I need to acknowledge the helpful staffs of the trade association members of the Global Wind Energy Council and European Wind Energy Association. I'm especially grateful to Randall Swisher, president of the American Wind Energy Association, and his staff Christine Real de Azua and Kathy Belyeu.

With little personal knowledge of print and digital photography, I relied on the expertise and generosity of my *American Shipper* colleague Keith Higginbotham. He used his precious time during the evenings and weekends to assist me with providing the best illustrations for this book.

Lastly and most importantly, I would like to thank my family and friends who quietly endured the numerous hours spent with my face buried in papers, books and the computer screen as I put this book together. My wife, Theresa, and children, Christopher and Elizabeth, were most encouraging and loving through the entire process.

# Chapter One
# Origins and History

It is uncertain which culture can lay claim to being the first to put wind to work on land, or for that matter how the idea got started in the first place. Was it a farmer's wife who was tired of the laborious chore of reducing grain into flour between two heavy stones for the family's daily bread? Did an ancient ruler decree his engineers to build a wind-powered device to centralize control of flour production for his people? Or was it an industrious sailor-turned-farmer who decided to harness the wind to improve his efficiency? We just don't know. Even China, which is often considered the developer of many early technologies, might not be the originator of large-scale windmills. What we can glean from fragmented accounts in history is that windmills have their strongest early roots in Persia, near the present day border of Pakistan and Afghanistan.

According to historians of mill history, a field known as "molinology," the oldest windmills in Persia were likely operated in a horizontal position, meaning that a set of light-weight sails mostly made from reeds were spaced within a cylinder, which in turn drove a vertical shaft leading to a set of millstones. Persia's early horizontal windmills are mentioned in a text from 950 AD referring to the tales of the three Banu Musa brothers of Baghdad between 850 and 870 AD. It's possible that the horizontal windmill operated in the region many years before the period referred to in the story. In 1872, Englishman Henry Walter Bellew visited the area and gave a good description of the mills. In much of the Middle East, windmills remained largely unknown until the late nineteenth century. Researchers uncovered an Arab drawing of a horizontal windmill from the thirteenth century, but have never found traces either in writing or through archeological investigations that Arabs actually constructed windmills. Since many early Arabs were nomadic peoples, there was perhaps little benefit to building windmills. About 1300, Mohammed Al Dimashqi provided a detailed description of a horizontal windmill in his book *Nukhbat-al-Dahr* (translated *Stories of the Centuries*). A drawing shows the windmill with millstones on the top floor and a horizontal wind wheel. The building surrounding the mill had several long vertical slots cut into the walls so that the wind could be channeled to the wind wheel.[1]

Ruins of a vertical axis horizontal windmill on the Greek island of Andros in the Aegean Sea in 1998. One mill of this type has been restored. *Courtesy of George Speis, Athens, Greece/The International Molinological Society, Congleton, United Kingdom.*

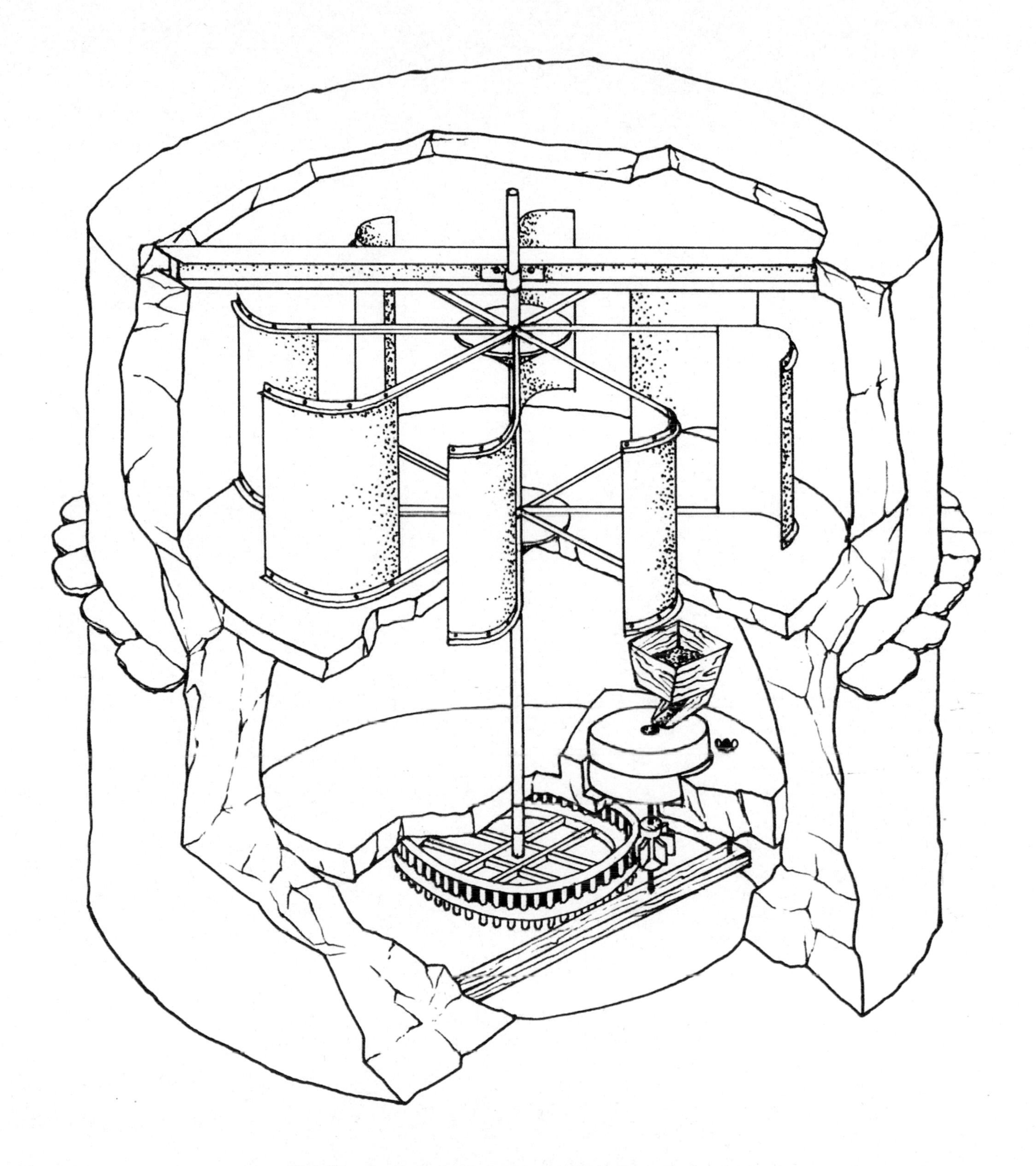

Drawing of a vertical axis horizontal windmill on the Greek island of Andros. *Courtesy of George Speis, Athens, Greece/The International Molinological Society, Congleton, United Kingdom.*

How the windmill first arrived in Europe is also up for debate. Some historians believe that windmills first appeared in Europe after the crusaders returned from Palestine. This story likely traces its origins to 1690 when A. Furestiere published his universal dictionary. Furestiere mentions that a crusader brought the windmill back with him to Europe. Amazingly, this story stuck and has been repeated by other historians and engineers over the centuries. However, this account is questionable at best, considering it's known that some Western Europeans made religious pilgrimages to Palestine as early as 950 AD. The first Crusade did not take place until 1096. If the windmills had been present in Palestine, why should these early visitors not have mentioned them? In 1190, Ambrosius wrote about the Arabs' surprise when the crusaders constructed a windmill in the city of Acre, now in Syria. It was considered the first windmill to be built in Syria.[2] The crusaders who built the windmill were under the command of Filips Augustus and Richard "the Lionheart." After 1190, no traces of any windmill activity can be found in the Middle East until the nineteenth century when some windmills were built in the region.

Many mid-twentieth century historians embraced the notion that early windmill developers applied the aspects of the Vitruvius water mill, which operated on a horizontal axis.[3] The oldest mention of a European windmill is 1183. In a document written by the Count of Flanders Filips van de Elzas (1168-1191), a decree was made that no one could build a windmill.[4] The document pertains to mills in Wormhout. This demonstrates that windmills were known in Flanders (in modern day Belgium) at the time. Researchers also found documents with references to windmills in Normandy thought to date to 1090, but they have since been determined to be from a later date. Changing dates on early documents was often done for tax reasons. In some cases, windmill dates were misinterpreted in translation of early documents in Latin. Some researchers translated the Latin word *molendinum* too often to windmill. The Count van de Elzas added the words *quod vento movetur* (translated "driven by the wind").[5]

Researchers have discovered other old documents which contain references to windmills. Early examples include: 1185 at Weedly, Sussex and Dinton, Buckinghamshire, both in the United Kingdom; 1192 at Cabtatre on the Somme in France (formerly part of Flanders, Belgium); and 1197 at Zonnebeke, near Ypres, Belgium. By the thirteenth century in the present day Belgian provinces of East and West Flanders there were more than 100 documented windmills.[6] Windmills were often built inside fortress walls throughout Europe to provide the inhabitants with flour during sieges.[7]

Post windmill illustrated in 1780 edition of French philosopher and writer Denis Diderot's *Encyclopédie, ou dictionaire raisonné des sciences, des arts et des métiers* (*Encyclopedia, or systematic dictionary of the sciences, arts and crafts*).

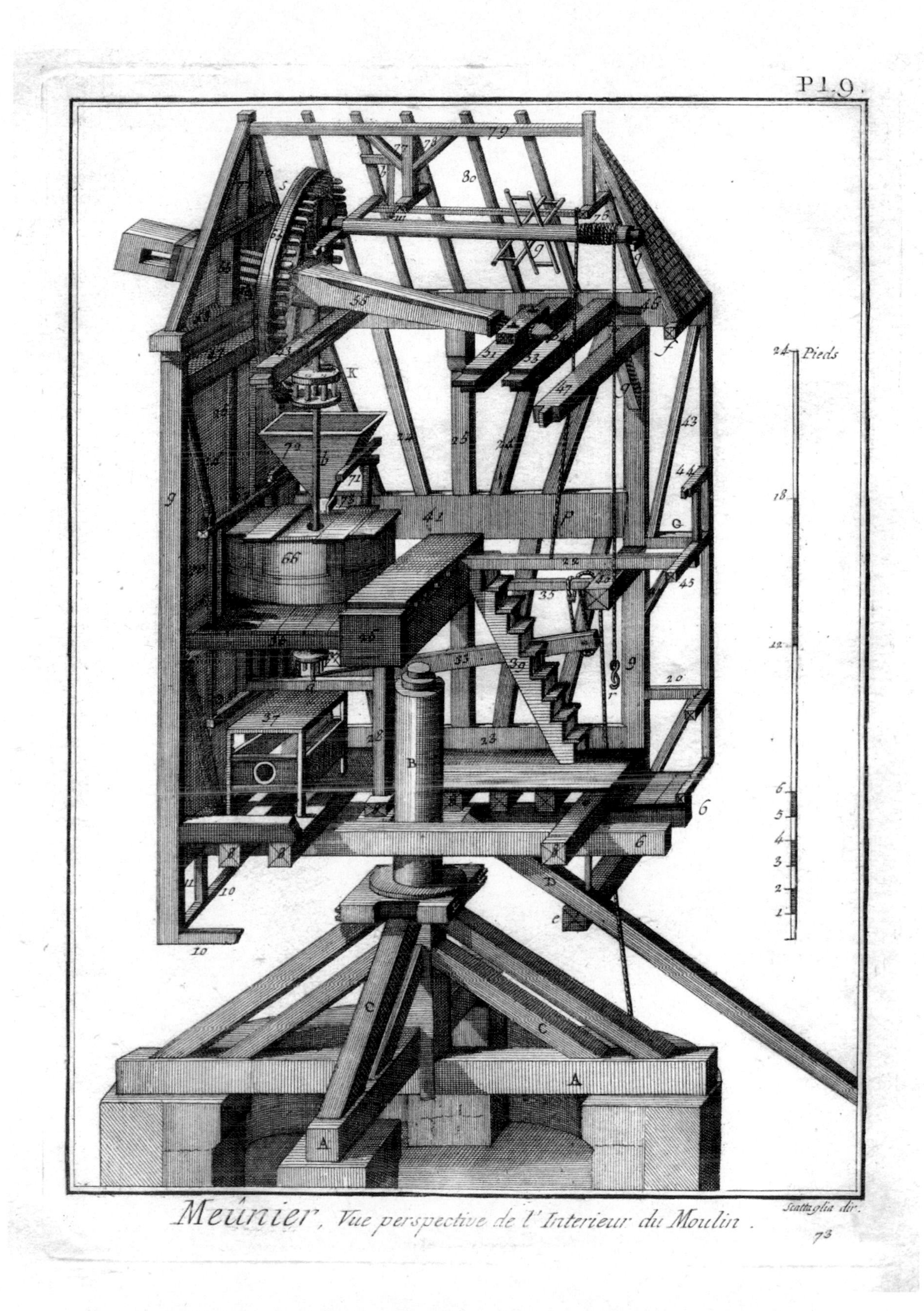


Post windmill interior illustrated in 1780 edition of French philosopher and writer Denis Diderot's *Encyclopédie, ou dictionaire raisonné des sciences, des arts et des métiers* (*Encyclopedia, or systematic dictionary of the sciences, arts and crafts*).

German soldiers stand under post windmill at St. Quentin, France. Postcard c. 1915.

The earliest representations of windmills appear to be the "post mill" variety, meaning that the mill housing was built upon a post. Windmill builders, known as millwrights, may have placed the original post mills upon large tree trunks or buried the posts in the ground.[8] By the late Middle Ages, millwrights suspended the posts on heavy timber trestles. The box-shaped mill housing, which was reached by a ladder, contained the equipment for making flour, such as wooden hoppers, millstones and gearing. The windmill's operation largely depended on the quality of sails and availability of wind. Early sails were generally a combination of wood lattice overlaid with cloth. The post allowed the miller to turn the windmill's housing and sails together into the wind. Sails were slightly angled or pitched to catch more wind. The cloth on the sails could be adjusted by the miller to help control the speed of the turning sails.[9] While inside the mill housing was a wind shaft brake mechanism, it was extremely dangerous to use it during strong winds because the sails could be suddenly snapped off and the intense friction between the brake and wind shaft could start a fire. Wind millers simply had to be vigilant at monitoring the wind and sail speed. A reason to have multiple pairs of sails on a windmill was because if one should need repair, the mill could be kept operational with at least one pair of sails. Most windmills had four sails, but some had as many as six and eight sails. The windmill building frenzy, which required long wooden beams suitable for sails, helped hasten deforestation in the United Kingdom by the fourteenth century.[10]

Stanley Freese, in his book *Windmills and Millwrighting*, noted that by the eighteenth and nineteenth centuries efficient English millwrights could build post mills relatively quickly. The Banner Mill at Quainton (Buckinghamshire), which was started on May 16, 1797, had the roof boards installed on the structure by June 16 that year. The sails were added 10 days later. The first sack of grain was ground into flour by July 20.[11] From start to finish, it took the millwright about nine weeks to build the Banner Mill. Freese attributed this efficiency to the layout of the work:

> The weightiest timbers for the new windmill would have been deposited conveniently around the mill yard, or in the mill field, so that they could readily be brought within reach of the lifting tackle by rolling them on logs, or turning them over with pinch-bars, or if necessary pulling them with a heavy rope coupled to a horse or a block-tackle. Everything was prepared in advance so far as main timbers and mill framing were concerned, with such items as sides of the mill-body probably being temporarily assembled and dismantled. The floor-boards and weather-boards, however, would be cut to length on the job, and the small fittings installed.[12]

In Europe, windmills were generally built starting in the spring and completed for operation by the end of the summer. A building time of six months was normal.

During the Middle Ages, other types of windmill structures became commonplace throughout Europe, such as stone and wood tower mills. These mills could be built taller and stronger than post mills, thus the cost to build them was generally more. While the towers were stationary, the caps through which the sails protruded could be rotated into the wind.

French tower windmill. Postcard c. 1910.

In 1258, a brick tower mill was constructed in Normandy. By 1294, a tower mill was noted at the castle in Dover, United Kingdom. Between 1250 and 1380, a number of brick tower mills were constructed in Flanders. In the region of what is today northern Germany and eastern Holland at least 30 tower mills were erected before 1400. It's estimated that seven builders needed two seasons – in winter the worked stopped – to build these mills. Some tower mills reached up to 100 feet tall with a wall thickness at the base of more than two feet. The oldest tower mills were cylindrical. The conical towers came later as millwrights perfected the tower mill's design. Short cylindrical towers spread across the landscape along the Mediterranean Sea. These mills were often simpler in design than their northern European counterparts. Broad cloth sails resembling those of sailing ships turned from stationary positions, taking advantage of the prevailing winds in the region.[13]

By the 1600s, another type of tower mill called a "smock" mill emerged in northern Europe. These wood-framed towers got their name because their appearance from a distance resembled a man's work smock. Smock mills were often constructed with octagonal sides that sloped out at the bottom. They were faced with boards and, in some cases, with thatch.[14]

Example of a Dutch smock windmill. Postcard c. 1920.

The Europeans took their expertise at building windmills to conquered lands overseas. In early America, the Dutch erected a windmill at the base of present day New York's Lower Manhattan. Reference to the windmill's sails is present in New York City's flag. In 1621, a post mill was built on the Flowerdew Hundred estate 20 miles outside the Jamestown settlement in Virginia. The windmill survived the Indian uprising of March 1622. The last reference to the Flowerdew windmill was in 1624 when the estate was sold by Sir George Yardley to Abraham Piersey. In the mid-1970s, English millwright Derek Ogden was commissioned to build an eighteenth century-style English post mill on the site of the original Flowerdew windmill. The replica windmill was opened to the public in April 1978.[15] Windmills were constructed along the U.S. East Coast from Florida to Maine and into Canada. In the mid-1800s, European-style windmills could be found throughout the Midwest in states such as North Dakota, Illinois, Kansas, Iowa and Texas. In 1865, H.F. Fischer built a smock mill in Elmhurst, Illinois that was 51 feet tall with sails spanning 74 feet across. With two run of millstones, the windmill had an output of about 40 barrels a day. It stopped grinding grain shortly after 1910.[16]

Windmills in Djibouti used for refining salt. Postcard c. 1905.

Tower windmill at Auckland, New Zealand. Postcard c. 1905.

Post windmill built on Maryland Eastern Shore in 1840. Postcard c. 1905.

Eight-sailed windmill near Newport, Rhode Island. Postcard c. 1910.

Windmill at Cedarburg, Wisconsin, built in 1873. Postcard c. 1940.

Smock windmill built at Reamsville, Kansas. Postcard c. 1905.

The mid-1700s ushered in an era of improved sail design for windmills. Two Englishmen – John Smeaton and Edmund Lee – individually made significant contributions to improvement of windmill sails. Smeaton studied numerous English and Dutch windmills to carry out his measurements of wind impact on sails. In a paper presented to the Royal Society in 1759, Smeaton explained that the power available from the wind is proportional to the cube of the wind speed, meaning that as the wind speed doubles the power from the wind increases eight times. Around the same time, Lee invented the fantail. This wooden multi-sailed device mounted at the back of the post mill or behind the cap of the tower or smock mill at a right angle to the main sails in the front. The fantail remains at the edge of the wind while the sails are turning. If the wind direction should change, the fantail would be set into motion and guide the windmill's sails back into the wind. This was a major labor-saving step for the miller, who could remain at work inside the windmill instead of stepping outside to manually turn the sails each time the wind changed direction. Post mills in Suffolk were nearly all tagged with fantails.[17] During this period, a number of inventors patented new types of windmill sails. One of the most interesting patent sails resembled a spring-loaded shutter, or Venetian blind. The shutters of the sails would open and close with the force of the wind, helping to control the speed that the sails turned.[18]

It was also during this period that windmills reached their serviceable peak to society. The English industrialists of the late 1700s played a significant role in expanding the application of windmill technology beyond just flour production. In the United Kingdom, the windmills became taller, the sails refined and the gearing modernized to drive more horsepower for heavier industrial processes, such as crushing oilseeds and oak bark (used for tanning leather), sawing wood, making fertilizer, shaping brass, and producing textiles. But the windmill's dominance as a power source quickly diminished in the 1800s with the increasing use of steam power.[19] People even lost their taste for stone ground flour preferring the whiter, more refined flour from steam-powered roller mills.

Another important use for windmills in Flanders and especially the lowlands of Holland was pumping water. Researchers found a document from 1316 that mentioned a windmill that pumped water from lower ground. This windmill was built in Drongen near Gent, Belgium.[20] About one hundred years later, the Dutch equivalent *poldermolen* helped claim marshy land for farming and residential living. There were several designs of drainage mills. From a distance, they resembled ordinary flour mills. Two popular designs were the small hollow post mill, or *wip* mill, and the large smock mill. Drainage windmills used internal Archimedean screws to lift water from the low wet ground into canal systems. They were commonly perched on dikes, giving them height to catch the sea breeze sweeping across the polders.

Bennet Flour Mill with 60-foot diameter sail wheel built in Bennet, Nebraska in 1874-75. *Courtesy of the Nebraska State Historical Society Photograph Collections*.

Post windmill in foreground shows fantail typical of the United Kingdom's Sussex area. Postcard c. 1910.

Smock windmill at Amagansett, Long Island, New York. Postcard c. 1915.

*People's Journal*, published in June 1854, offers notice of a $30 water-pumping windmill with a 10-foot sail diameter.

Starting in the mid-1800s, the Americans began developing their own style of windmill with an emphasis on pumping water. The earliest American windmills employed wooden sails that tied into a geared mechanism which sat on top of a tower. When the wind hit the sails, the gears activated to provide the up and down motion for the pump rod which lifted water from inside the ground to the surface. The water could be easily stored in tanks for farming and household purposes. Besides agriculture, these windmills became important to the railroads whose locomotives routinely required water for their coal-fired steam engines. By the 1870s, these windmills were found throughout the country.[21]

Daniel Halladay, a mechanic from Connecticut, is largely considered by historians to be the inventor of the so-called "self-regulating" windmill to pump water and grind grain. Halladay's windmill employed a tail that automatically turned the wind-wheel face, or sails, into the wind. The Halladay Wind Mill Company began supplying its product to the inhabitants of the thirsty prairie in 1857 through the U.S. Wind Engine and Pump Company in Chicago.[22] It wasn't long before other windmill designers emerged on the market. There were notable brands, such as Eclipse, Heller-Aller, Perkins, Star, Dempster, Fairbury and Aermotor. The wind-wheels ranged from six feet to eight feet in diameter at the low end to enormous sizes, depending on the well depth and volume of water sought, or other mechanical functions, such as grinding grain and sawing timber.

American windmill designers delighted in showing off their work on enormously tall towers. At St. James on Long Island, New York in 1894, Andrew J. Corcoran erected a 22.5-foot diameter wind-wheel on top of a 150-foot tall tower. The windmill required 6,000 feet of pipe between the pump and the reservoir below ground. The water poured into a 65,000-gallon holding area, which could be filled by the windmill in two days.[23] The same year, at the World's Columbian Exposition, Aermotor placed a 16-foot diameter wind-wheel on top of a 150-foot tower that was anchored to the base of a Dutch-style windmill. The May 12, 1894 *Scientific American* reported: "This windmill towered above all competitors at the Fair, and around the lower structure was a balcony nearly 150 feet in circumference, from which an impressive idea of the height of the tower was obtainable."[24] Wooden wind-wheels remained popular until around World War I. Aermotor was the first windmill manufacturer to construct an all-steel windmill. All-steel construction became the norm among windmill manufacturers by the 1920s.[25] Some American windmill manufacturers promised much more performance than their windmills could actually deliver, leading to their demise in the market. At best, most of these windmills pulled enough water to irrigate vegetable gardens and small orchards, and fill livestock troughs.[26] The U.S. Patent Office received thousands of patent applications covering the gamut of windmill design. Dr. T. Lindsay Baker, one of the country's prominent American windmill historians, found that the U.S. government issued more than 3,500 patents for wind machines between the 1790s and 1950.[27] While these windmills were primarily used to pump water, some were used to grind grain and saw wood. The American-style windmill could also be found throughout the world by the late 1800s and early 1900s. Australia, for example, developed a healthy water-pumping windmill manufacturing industry of its own. Manufacturers in Germany came up with American-style wind pumpers after witnessing these devices in action at the 1876 Centennial Exhibition.[28]

The 1936 catalogue for the Heller-Aller "Baker" water-pumping windmill. The Napoleon, Ohio-based company ceased operation in the early 1990s.

A restored Halladay "Standard" Windmill. *Courtesy of the American Wind Power Center, Lubbock, Texas.*

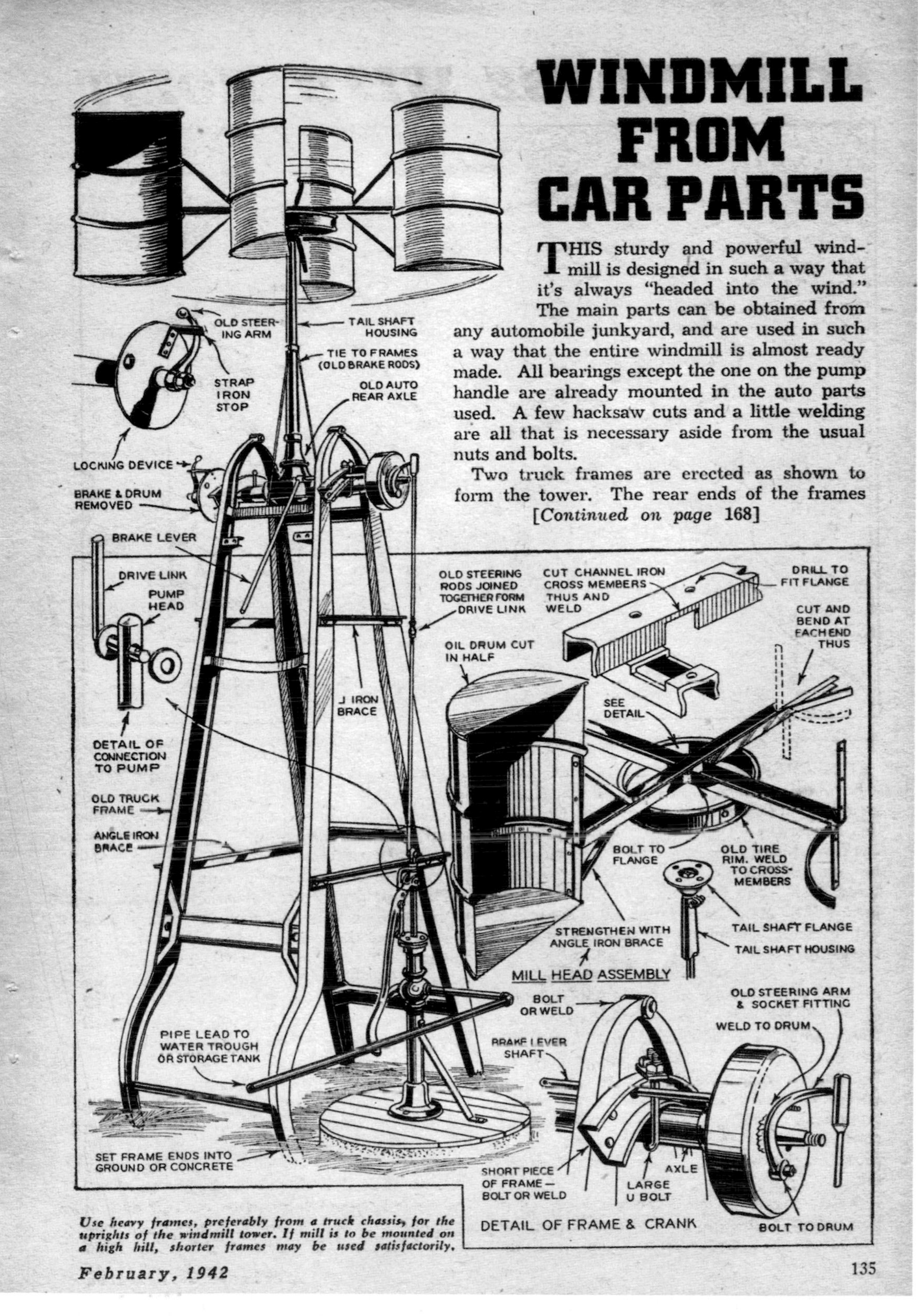

# WINDMILL FROM CAR PARTS

THIS sturdy and powerful windmill is designed in such a way that it's always "headed into the wind." The main parts can be obtained from any automobile junkyard, and are used in such a way that the entire windmill is almost ready made. All bearings except the one on the pump handle are already mounted in the auto parts used. A few hacksaw cuts and a little welding are all that is necessary aside from the usual nuts and bolts.

Two truck frames are erected as shown to form the tower. The rear ends of the frames

[*Continued on page* 168]



*Use heavy frames, preferably from a truck chassis, for the uprights of the windmill tower. If mill is to be mounted on a high hill, shorter frames may be used satisfactorily.*



The February 1942 issue of *Mechanix Illustrated* tells the reader how to build a water-pumping windmill with car parts.

of the wind when working. The windmill plant at Wittkeil, in Schleswig, has demonstrated certain facts that carry out the foregoing statements. In this case the windmill has an enormous wind surface. The diameter of the windmill is about 40 feet and an effective wind space of nearly 1,000 square feet is presented to the wind. The wind develops 30 horse-power with a normal speed of eleven revolutions per minute. It operates a shunt dynamo that makes 700 revolutions per minute, and develops 160 volts and 120 amperes. This full load is developed when the wind is blowing at the rate of about 8 miles per hour. The windmill furnishes electricity to light the town of Wittkeil, and small motors and lamps are connected to the storage battery, which maintains a voltage of 110. The battery has a capacity of 66,000 watt-hours. This plant has proved so satisfactory that it is being enlarged, and as a permanent lighting station it is likely to prove of unusual importance in the development of modern electricity by windmills.[7]

At the start of the 1900s, there were several successful experiments with wind-generated electricity in Germany. Two of these experiments proved successful in Hamburg and near Leipsic. The *Scientific American* reported in 1904 that the windmills were "strongly built, and designed to take the wind at any angle. The regulation of the motor is obtained by means of an automatic switch, which cuts out the battery when the wind falls to a low pressure."[8]

However, the unsung pioneers of wind-driven electric plants were most likely farmers and eccentrics who wanted to improve their lot in life. They became aware of the then wonders of electric lighting, refrigeration, heating, and battery charging during their occasional trips into town and from reading the newspapers and magazines of the day.[9] Farmers were generally skilled in basic mechanics and knew how to make do with what money and resources they had available. In the Great Plains, one of the greatest resources to farmers was the wind. Some farmers surmised that if they could use the wind for mechanical activities, such as water pumping and making feed, then they should be able to harness it to generate electricity.

Electrical engineer Putnam A. Bates put a name to one of these otherwise unassuming wind energy pioneers in a 1912 article published in the *Scientific American*. Bates described how J.F. Forest constructed an electricity-producing wind plant on his Poynette, Wisconsin farm for about $250. Forest got to thinking about electricity generation after watching his water-pumping windmill turning in the breeze. He reasoned if the windmill's armature could pump water from the ground, then it should be able to drive the armature of a dynamo to produce electricity. Forest hitched a 12-foot diameter wind wheel to a vertical shaft which extended down through the windmill tower to a series of pulleys, bevel gears and a set of grinder rings. He built his electric plant on the second floor of a barn. The windmill drove a dynamo with a 0.21 kilowatt capacity. In an adjoining barn was the 14-cell storage battery unit. The electric plant provided power for 24 tungsten lamps scattered throughout the house and farm buildings. Forest also used the same windmill to turn a drill press, grindstone, corn sheller, beehive saw, washing machine, grain elevator and feed grinder.[10]

Poul la Cour's first test turbine built at the Folks High School in Askov, Denmark in 1891. *Courtesy of The Poul la Cour Foundation, Middelfart, Denmark.*

3. Hæfte. Maj 1904.

TIDSSKRIFT FOR VIND-ELEKTRISITET

UDGIVET AF POUL LA COUR

GYLDENDALSKE BOGHANDEL · NORDISK FORLAG

In 1904, Poul la Cour formed the Danish Wind Electric Society (Dansk Vind Elektrisitets Selskab) and published a journal about wind-generated electricity. *Courtesy of The Poul la Cour Foundation, Middelfart, Denmark.*

In 1917, Oliver P. Fritchle, who had been manufacturing electric-battery-powered automobiles since 1904, began selling wind-powered electric generators under the name "Fritchle Wind-Electric Plants." The generator sat on top of the windmill tower and power was carried through wires to a control panel and storage batteries to power lights, radios, and other equipment. While the concept was intriguing, it wasn't a commercial success. The Woodmanse Manufacturing Company, which produced Fritchle's wind plants, sold only a modest number these machines, mostly to customers on the Great Plains, and a few as far away as Brazil and Argentina. The Fritchle Wind-Electric Plants were discontinued in 1923.[11] Other American windmill manufacturers tried to tap the electric-generation market but with minimal success due to technical limitations.

Battery-powered automobile builder Oliver P. Fritchle and one of his wind-electric generating plants at Warminster, Colorado (c. 1920). *Courtesy of the Stephen H. Hart Library, Colorado Historical Society, Denver, Colorado.*

The combination post/tower electric-generating windmill *Merlaan* (c. 2007) erected by Alfred Ronse at Gistel, Belgium in 1934. *Courtesy of Levende Molens, Ekeren, Belgium*.

While newer, more efficient turbine designs surfaced in the early 1900s, there were still individuals who didn't give up on the traditional windmill designs to generate electricity. One of those individuals was a Belgian politician and molinologist named Alfred Ronse, who in 1934 built a wind-powered electric plant in the village of Gistel. Ronse was particularly keen on experimenting with the Dekker sail design and special bearings for wind shafts. His plan for the windmill was unique. He constructed a brick tower mill and then placed a wood post mill on top. The post mill had four Dekker sails, which had a rotating diameter of 56 feet (17 meters), and housed the bearings. The reason Ronse chose this design remains unclear. The windmill, named "Meerlaan," lacked a fantail winding system and required a miller on site during operation. A steel shaft, which turned by the action of the sails, ran through the post mill and into the tower, linking to the dynamo. To obtain the correct speed on the dynamo – 1,450 rpm for 12 kilowatts production – a considerable transmission system was required. The wind shaft, which turned at 15 rpm, was stepped up to the required 1,450 rpm. Ronse's design largely compensated for changing wind shaft speeds, resulting from fluctuating winds, by using a stage-free drive system. Essentially, the governor took into account the actual speed of the wind shaft and shifted a belt along conical pulleys. In a later stage, a smaller dynamo was added to the windmill. The system to drive this dynamo was far less complex than initial larger dynamo. The direct current generated from the dynamos was stored in a battery bank.[12] The Meerlaan still stands as a testament to Ronse's noble effort.

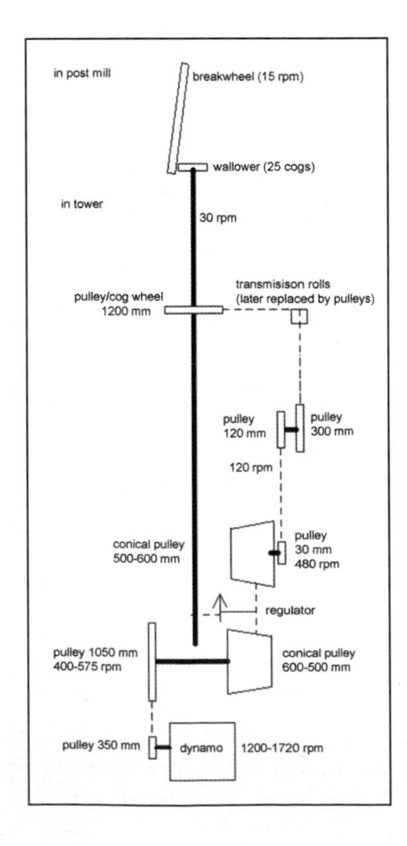


Schematic showing how electricity was generated by the *Meerlaan* windmill. *Courtesy of Frans Brouwers, Levende Molens, Ekeren, Belgium*.

# Chapter Three
# Small Wind Power

Wind turbine developments for rural household and farm use in the early 1900s remained largely a tinker's domain. This began to change shortly after World War I when some industrious individuals with backgrounds in basic aerodynamics realized the financial potential for standardizing wind turbines to feed America's increasing desire for electrically powered appliances. Farmers, who were often located far away from the electric power grid of the cities and towns, took their 6-volt radio batteries to town for charging at auto repair shops. The trek required time away from work and often a wait of several days before the recharged batteries were available for pickup.[1] In 1916, the Dayton Engineering Laboratories Company (Delco) introduced a farm electric plant consisting of a single-cylinder I-C engine/generator set and battery. The device was used to charge a battery and shut off automatically when it was charged. The engine kicked back on when the battery's power decreased to a certain level. The charge process took about four to six hours. The battery could then be used to provide power to lights, appliances, motors and well pumps. Other noted manufacturers of gensets at the time were Onan (known today as Cummins), Kohler and Fairbanks-Morse. But some farmers found less expensive ways to charge their batteries by attaching simple windmill blades to automobile alternators. In the 1920s and 1930s, some inventive entrepreneurs realized a market existed in rural America for standardized wind energy plants.[2]

One of the earliest and most notable successes with mass-produced wind turbines was the two-bladed Wincharger. In the early 1930s, brothers John and Gerhard Albers of Cherokee, Iowa began building small wind turbines to charge radio batteries. Their neighbors took notice of their work and began placing orders for their wind turbines. By 1935, the Albers started the Wincharger Corporation and opened an assembly plant in Sioux City, Iowa. That same year, radio manufacturer Zenith Corporation acquired a controlling interest in Wincharger and began jointly marketing their products to consumers. Zenith bought out the rest of Wincharger in 1937 and expanded the power capacity of the wind plants to include 500 watt, 750 watt, 1,000 watt and 1,200 watt generators. Wincharger also sold towers, battery sets, motors, and appliances, such as sewing machines, vacuum cleaners, and dairy equipment.[3] In 1937, Zenith launched an aggressive marketing campaign. According to company literature at the time, Zenith claimed to sell more than 750,000 units by 1938.[4]

The difference between these wind turbines and earlier models is in the design of the wind wheel. Earlier power-generating wind wheels used multi-bladed patterns. These wind wheels had high torque to pump water and were good at operating at low wind speeds, but the problem was that too much of the wind's useful power got absorbed by the blades themselves, known as "wind congestion."[5] Wind power developers began taking interest in the sleeker propeller design used on early airplanes. The propeller allowed wind to pass through it helping to give an airplane lift. These types of propellers, used in the context of wind-based electricity generation, operate efficiently in winds ranging from 8 to 25 mph. The Alber brothers, and other developers of the time, recognized this quality in the propeller's shape and similarly used the lift from the wind to power their generators.

A 1930s marketing campaign launched by Zenith to promote its combination Wincharger and radio package.

The 2.5 kilowatt Jacobs Wind Electric Plant constructed at Adm. Richard E. Byrd's "Little America" camp in Antarctica in 1933, which remained operational until 1955. *Courtesy of Craig Toepfer, Ann Arbor, Michigan.*

While both Jacobs brothers were inventive, Marcellus was the marketing persona behind the company's success. Thousands of Jacobs Model 45s were sold throughout the world. The Jacobs prided themselves on the durability of their wind turbines. They operated successfully in punishing climates, such as Alaska, Canada, Finland, northwestern United States, and at the joint U.S.-British weather station at Eureka in the Arctic Circle.[7] In 1933, Admiral Richard E. Byrd used a Jacobs turbine to provide power to his Little America camp at the South Pole. According to Byrd's son, who visited the site in 1955, the Jacobs turbine was still in operational condition. Richard E. Byrd, Jr. wrote to Marcellus Jacobs that same year stating: "I thought it might interest you to know that the wind generator installed... at the original Little America, was still intact this year after a quarter of a century... The blades were still turning in the breeze (and) show little sign of weathering. Much of the paint is intact."[8] The Jacobs brothers coated their spruce wood propellers with aluminum paint, which better resisted frost and ice formation than other more common varnishes and coatings used by the industry. The brothers also found a remedy to damaging lightening strikes by installing dual sets of heavy grounding brushes on the armature staff. With the added use of a large capacity oil-filled condenser connected across the generator brushes and frame, the Jacobs practically eliminated any damage to the generators from lightening. In addition, the Jacobs wind turbine's effectiveness led it to be used for some industrial applications. In 1936, special generators were designed for the cathodic protection of underground steel pipe lines used by

Jacobs Wind Electric Plant proved it could endure the harshest winter conditions in Antarctica. Ice buildup at "Little America" camp reached 70 feet against the turbine's tower by 1947. *Courtesy of Craig Toepfer, Ann Arbor, Michigan.*

the oil industry in North and South America and the Middle East.[9]

During the 1940s and 1950s, a 2,500-watt, 32-volt Jacobs wind plant could be purchased from the factory for $490, excluding the cost of the tower and batteries, which were often bought locally. Jacobs offered a complete line of towers and batteries to customers for a total wind plant cost of $1,025. "Installation cost requires only the labour of two men for two days and a small amount of cement to put into the anchor holes when the tower is built. No special equipment or training is necessary. We have shipped hundreds of plants to most countries with not a single request for additional information to enable them to erect the plant," Marcellus Jacobs wrote at the time.[10] So confident were the Jacobs in their turbines that they offered customers a five-year guarantee on parts. The biggest maintenance cost was the replacement of the storage battery, which needed to be changed on a 10-year basis at a cost of $36 per year.[11] Other repairs were generally less than $5 per year. Between 1930 and 1956, it's estimated that Jacobs built and sold more than $50 million worth in turbines.[12]

There were several dozen successful wind turbine manufacturers to emerge on the market in the mid-1930s, with many located in the Midwest. They included the likes of Winpower, Parris-Dunn, Air Electric Machine Company, Nelson Electric and Lejay. Numerous tinkers promoted their efforts to build low-cost homemade wind engines, but often with more hype than genuine success. The Europeans developed their own small wind turbines using propeller-type blades. For example, the Kumme system developed in Germany in the 1920s used four-bladed rotors of up to 60-feet in diameter. The turbine used two smaller wind wheels to perform the work of a vane to hold the rotor in the wind.[13] However, the United States remained the dominant global manufacturer of small wind power plants.

This first period of small electric-producing wind turbines in the United States was headed for a slow demise. In 1938, Congress passed the Rural Electrification Act, which promised to extend electric power lines to rural Americans. By the mid-1950s, few remote farms and towns were without electric power. Farmers with wind turbines were often forced to remove or destroy their wind turbines in order for the power companies to hook up to their properties.[14] The Jacobs Wind Electric Company bought other companies for their copper allotments during World War II so it could keep manufacturing its wind plants, but even it was forced to close shop in 1956. The boisterous Marcellus Jacobs would pursue other interests before stepping back into the small wind energy business nearly 20 years later.

Interest in small turbines reemerged after the 1973 Arab oil embargo. This event emphasized America's dependence on Middle Eastern oil. Panic quickly set in when rumors circulated that oil prices could reach as high as $100 a barrel. At the same time, more Americans became concerned about pollution. Some even divorced themselves from all connections to the national power grid. Suddenly, a market for reconditioned Jacobs and Winchargers took hold. Surviving turbines, laid up in old barns and sheds, were dusted off, reconditioned and put back into operation. Entrepreneurs surfaced with plans for the average person to build small wind turbines from materials found around the garage and in junk piles – some successful, many not.

Even the federal government's interest in small wind turbines at the time stimulated renewed industry optimism. In the early 1980s, the Bonneville Power Administration, an agency of the Department of Energy based in Portland, Oregon, tested a range of small wind turbines in the region to determine whether they could serve as supplemental power generation sources to the hydroelectric dams along the Columbia River.[15] The Department of Agriculture's Agricultural Research Service in Bushland, Texas became a test bed for many types of small turbines coming onto the market. The agency's primary focus was on the application of turbine-generated electricity for powering irrigation systems. At the time, the Agricultural Research Service and the West Texas University's Alternative Energy Institute estimated the irrigation systems for agricultural purposes required 90 billion kilowatts per year. More than 473,000 on-farm pumping plants were in operation in the United States in the early 1980s, and about half were in service in the Great Plains, a region known for its abundant wind. The Agricultural Research Service tests demonstrated the flaws in design and operational shortcomings when it came to specific applications, such as irrigation and utility grid connections.[16]

LEJAY
WIND ELECTRIC
EQUIPMENT

MANUFACTURERS OF
WIND ELECTRIC EQUIPMENT, Etc.
DEAL DIRECT
SAVE THE DIFFERENCE
WE SHIP ALL OVER THE WORLD. REMIT BY INTERNATIONAL MONEY ORDER

- ELECTRIC BATTERY RAZORS
- WIRE
- GENERATORS
- PROPELLORS
- WELDERS
- ELECTRIC FENCE CONTROLS
- LEJAY MANUAL OF 100 GENERATOR CHANGES

CATALOGUE No. 26
Winter 1939 — Spring 1940

LEJAY ELECTRIC WAY

Winter 1939/Spring 1940 catalogue (No. 26) of Minneapolis, Minnesota-based wind electric plant manufacturer Lejay.

Earth Energy Systems, formerly Jacobs, sold a 20 kilowatt wind turbine in the mid-1980s. *Courtesy of Craig Toepfer, Ann Arbor, Michigan*.

Former market leaders, such as Jacobs and Winco, re-entered the business with "new and improved" versions of their double and triple bladed turbines. Marcellus Jacobs joined his son Paul in 1972 to revive the Jacobs Wind Electric Company. Computer manufacturer Control Data Corporation of Minneapolis, Minnesota became a partner in the company. Jacobs opened a new production plant for his turbines at Plymouth, Minnesota in 1980. The company made a few acquisitions. In 1984, Jacobs bought wind farm developer Renewable Energy Ventures in Enrico, California, followed a year later with the purchase of Dyna Technology and its 250,000-square-foot manufacturing plant in LeCenter, Minnesota, which was the maker of Winco (formerly Wincharger) wind turbines and engine generators. The Jacobs company was renamed Earth Energy Systems with the Jacobs names still associated with the wind turbines. The company successfully sold a 20-kilowatt wind turbine. Jacobs himself bowed out of the day-to-day operations by the mid-1980s, but remained active as a consultant to the company. He encouraged the development of new applications for small wind turbines. For example, the company's engineers developed a modular hybrid power system for stand-alone operations in remote areas. The system consisted of a 20-kilowatt Jacobs wind generator, tower, battery set, and microprocessor controls, including load management, distribution/protection, and power conditioning equipment. The system could be integrated with other power generation equipment, such as photovoltaic solar panels and gas motors. An individual system was contained and transported to a site in a 20-foot maritime container. In 1985, Earth Energy Systems also retooled the 1,200-watt four-bladed Wincharger from the 1940s and developed a small Winco hybrid system around this machine. The new four-bladed Wincharger was rated at 1,200 watts and 1,400 watts and was available with the company's 30 foot-to-120 foot guyed towers, with freestanding towers available from partner suppliers. The system included a small Winco generator set – available in gas, diesel, natural or propane modes; a 1,200-watt Winco inverter; and combined with a deep-cycle lead-acid battery set and solar panels for a cost of $5,000. But the company's ambitious projects came to a standstill in the mid-1980s when Control Data suffered financial setbacks and eventually folded.[17]

Marcellus Jacobs commonly referred to newcomers to the industry as "young whippersnappers." One of those whippersnappers was the Bergeys, a father-son team who put their mark on the wind turbine business in the mid-1970s and early 1980s. Founder Karl Bergey had served as assistant chief engineer at the Piper Development Center in Vero Beach, Florida from 1957 to 1968, and is widely credited for design of the Piper Cherokee aircraft. He next joined the engineering faculty at the University of Oklahoma and began researching small wind turbine designs. Together with his son Michael, who obtained a mechanical engineering degree, they formed the Bergey Windpower Company with $40,000 in 1977. Michael Bergey, who now serves as the company's president, recalled in an interview with the author his father asking him "How little can you live on?" just before starting the company. The small wind market at the time was still negligible, but optimism was in the air.[18]

The Bergey Windpower Company spent its first two years developing a horizontal axis wind turbine system. They tested a number of new technologies at the time, such as special airfoils, passive blade pitching, rotor speed controls, and low-speed permanent magnet alternators. The Bergey philosophy was "simple is better." With fewer moving parts, other than the turbine blades, maintenance diminishes substantially. Even the venerable Jacobs required routine repairs, such as bushing replacements. The Bergeys had the benefit of access to lighter fiberglass blades, permanent magnets and better electronics to apply to their wind turbine systems. However, numerous tests were conducted before a product was ready for market. "You have to build a good pile of wreckage before you come up with something that works," Michael Bergey said. The first production models, namely the 1 kilowatt BWC 1000, still suffered problems until the kinks were worked out. Ultimately, they reached success and the company produced more than 600 BWC 1000 units between 1980 and 1990, before the system was upgraded to a 1.5 kilowatt unit. The BWC 1000 was sold in the United States and to more than 50 countries. Over the years, the Bergeys developed larger wind turbines. In 1983, the company introduced the 10 kilowatt BWC EXCEL, which became popular among farmers and for overseas rural village electrification projects. According to the company, more than 900 BWC EXCEL units have been installed and have clocked more than 60 million hours in operation.[19]

In 1978, Congress passed the Public Utility Regulatory Policies Act (PURPA), which required utilities to buy excess electric power at reasonable market rates from small wind turbine owners. The law also spared small turbine operators from compliance with the Federal Powers Act, which governed the operations of public utilities. In a 1981 article, Paul Gipe, a recognized authority on small wind energy and proponent of beneficial government regulations for the wind energy industry, called PURPA "downright revolutionary," because it encourages "decentralized energy investment" and "alters our

in the slightest breeze and are capable of charging batteries for household appliances. One of these units was installed on the roof of Hong Kong's Sea School in April.[32] Japanese vessel operator Mitsui O.S.K. Lines (MOL) began testing a new wind power generator aboard its wood chip carrier, *Taiho Maru*, in 2004. MOL sought the turbine as an internal initiative to introduce environmentally friendly technologies to its vessels. Tokai University Research Institute of Science and Technology and Nishishiba Electric worked together with MOL to develop and install an onboard, straight-wing, vertically axis-type wind power generator. The generator's compact shape and omni-directional rotation allow it to generate power no matter which direction the wind blows. The device was installed on the *Taiho Maru*'s bridge, where it received the strongest winds.[33] While the test was considered successful, MOL has no immediate plans at this time to install turbines on its other vessels. MOL and Tokai University will continue to explore whether the turbine's electrical output could be sufficiently stabilized to power sensitive navigation and other ship-critical systems.[34]

Established small wind turbine manufacturers remain weary of poor performing units permeating the market and tarnishing the industry's reputation. Michael Bergey, chairman of the AWEA's Small Wind Turbine Committee, has seen this happen all too often during the boom times in the industry. He refers to these developers as "bozos" (those who are clueless about physics and engineering) and "shysters" (those who are aware that their claims are bogus and don't care).[35] In 2007, the National Renewable Energy Laboratory announced its offer to perform independent tests at its National Wind Technology Center on up to eight commercially available small wind turbines. This will allow the turbine manufacturers the opportunity to earn a certification granted by an independent certification body. The Small Wind Certification Corporation, which will form the independent certification body, is a non-profit organization led by the Interstate Renewable Energy Council, with support from the Department of Energy's Wind and Hydropower Technologies Program, AWEA, state energy offices and wind turbine manufacturers. The National Renewable Energy Laboratory will evaluate turbines based on testing protocols, which are defined by the independent certification body, and by the AWEA's standard for small wind turbines. Tests will include duration, power performance, acoustic noise emissions, safety and function, and power quality. The results of the tests will be made available on the National Renewable Energy Laboratory's Website.[36]

University of Hong Kong engineers in early 2007 designed a micro-wind system with interconnected 10-inch plastic gear wheels for use in urban and rural settings. *Courtesy of Motorwave Group, Hong Kong*.

Japan's Tokai University Research Institute of Science and Technology and Nishishiba Electric worked together with Mitsui O.S.K. Lines in 2004 to develop and test a straight-wing, vertical axis-type wind generator on board one of the ocean carrier's ships. *Courtesy of Mitsui O.S.K. Lines, Tokyo, Japan.*

# Chapter Four
# Scaling Up

After World War I, some scientists and engineers believed energy for the wider community's use could be tapped from the wind. It simply required larger machines to efficiently harvest the energy. In the early 1920s, two Germans separately attempted to put their large-scale wind turbine ideas to work. Anton Flettner, an inventor known for his wind-powered electric rotor ship which successfully crossed the Atlantic Ocean in 1925, drafted plans to build a wind turbine with 300-foot diameter blades on top of a 650-foot tower. But instead of using a single generator for the turbine, Flettner proposed to mount a smaller high-speed turbine on the tip of each propeller arm. Flettner claimed that the rotation of the large wheel would multiply the velocity of the small turbines 10 times and that this would help to regulate the electric power output.[1] In 1925, Hermann Honnef, a professor in Berlin, wanted to build a 1,000-foot tower to support five 250-foot diameter rotors, which he said could generate up to 50 megawatts of power a year.[2] Only the so-called Kumme design turbine was put to work in Germany in the 1920s. It consisted of six pitched blades. While the turbine guided itself into the wind by using the fantail design of early European mechanical windmills, its generator was on the ground and proved uneconomical.[3] Other European countries witnessed the launch of other large-scale wind machines during this period, such as the two-bladed turbine constructed by Compagnie Electro-Mecanique in 1929 at Bourget, France.

In 1931, the Soviet Union built a more practical large-scale wind turbine on a bluff at Balaclava near Yalta to provide supplemental electricity to a peat-burning 20,000 kilowatt steam power plant about 20 miles away at Savastopol. The turbine's blades measured 100 feet in diameter. The automatically pitched blades were turned into the wind by a rigid strut system that connected to a circular track surrounding the turbine's base. The turbine's main gears were reportedly made of wood and the blades' skins were wrapped in roofing metal. It was estimated that with average 15 mph winds the turbine could generate about 279,000 kilowatts a year. The Soviet Union's ambitious plans to further exploit its wind energy were set aside at the start of World War II.[4]

Also during the early twentieth century, some engineers broke away from traditional designs with the vertical wind turbine blades attached to a horizontal axis. In 1925, Frenchman Georges Jean Marie Darrieus patented a wind turbine with blades that resembled an upright eggbeater, a vertical axis concept that harkened back to the early Persian windmills. At about the same time, S.J. Savonius of Finland developed a rotor design that used a vertically split cylinder, separated slightly to cup the wind as it turns.[5] Inventor Julius D. Madaras developed a wind power scheme that consisted of a series of large cylinders – each measuring 90-foot high and 22-foot in diameter. Madaras erected a full-scale test cylinder at Burlington, New Jersey in October 1933.[6] The cylinder turned on a vertical axis to generate a large "lateral force," known as the "Magnus Effect." It was Madaras's goal to mount these cylinders to a string of cars carrying generators. When the wind hit the cylinders, the cars would start moving along a half-mile diameter track. The cars' wheels would drive the generators much like a streetcar going down hill. According to Madaras, 40 rotors would be required to generate 50,000 kilowatts. Madaras claimed at the time that the cost of his power unit would be far less than to build the average steam or hydroelectric plant.[7]

These early efforts became the inspiration for the work of Palmer Cosslett Putnam, a one-time geologist, American flyer for Britain during World War I, and former president of New York publisher G.P. Putnam's Sons. Putnam got his idea to build a large-scale turbine in the mid-1930s when he considered using wind to generate power for his Cape Cod home.[8] He studied the available information of the period, particularly the technical specifications of the Soviet and Honnef designs. He concluded that the Soviet model suffered from "low efficiency, crude regulation and yaw control, high weight per kilowatt, and induction generation," while the Honnef concept "exaggerated the importance of height." Most importantly, none of the previously published designs were economically sound.[9] In early 1935, Putnam presented his idea to Vannevar Bush, dean of engineering at the Massachusetts Institute of Technology, who in turn referred him to Tom Knight, General Electric's vice president in Boston. Within 10 minutes of their meeting, as the story goes, Knight offered Putnam office space and engineering help to build his large wind turbine. Knight was instrumental in getting the New England Public Service Company to buy sites for the experimental turbine.

In 1939, Knight helped Putnam sell his idea to the S. Morgan Smith Company, a maker of water turbines.[10] The biggest challenge for Putnam was to design a generator with accurate speed regulation to provide consistent power despite constantly changing wind speeds.[11]

Putnam's determination to build the turbine never waned, even in the face of the Second World War. The project was promoted for its energy independence during times of war and national emergency. A press release from the time stated that "with an eye to national defense, they also say that a series of such wind turbines, distributed through the hills, would be less vulnerable to air attack than equivalent generating capacity in a single station."[12] It was determined that Grandpa's Knob, a 2,000-foot summit located between Castleton and West Rutland, Vermont, was a suitable site to test the giant turbine, and construction began in earnest during the summer of 1941.

Palmer Cosslett Putnam's 1,250-kilowatt utility-scale turbine erected at Grandpa's Knob, Vermont, in 1941. *Courtesy of Paul Gipe, Tehachapi, California.*

British government stopped funding wind turbine research by 1960, opting to explore the potential of nuclear energy.[29]

Other European countries erected large, mostly experimental wind turbines of their own during the 1950s and early 1960s. In 1953, Honnef of Germany received backing to build a wind turbine of his design, although much smaller than he originally planned. The town of Salach in southern Germany raised enough funds to build a Honnef-tower with turbine blades spanning 130 feet in diameter to supplement power to its coal-burning power plant.[30] The French Bureau d'Etudes Scientifiques et Techniques developed an 800 kilowatt turbine for the national electricity authority. In 1958, a three-bladed prototype, with its 100-foot rotor, was set up southwest of Paris near Nogent Le Roi. The turbine's conical tower was built upon a four-legged "Eiffel-like" tower attached to the ground.[31] When new blades were installed in 1962, one broke off destroying the turbine's hub. The machine was dismantled in 1965.[32] The French next installed two smaller three-bladed turbines in southern France near St. Remy des Landes in 1963. One turbine had a rotor diameter measuring 70 feet, and generated up to 130 kilowatts in a 28 mph wind. The second turbine's rotor diameter measured 100 feet and produced as much as 300 kilowatts in winds of 37 mph.[33] The bigger of the two French turbines experienced mechanical failure in 1964 and the government decided to abandon its wind program in 1966.[34]

Simply put, wind turbine-produced power struggled to compete with traditional power sources in the 1950s and 1960s. George A. Whetstone, a professor in the Texas Technological College's Department of Civil Engineering, offered a glimmer of hope for wind energy and in 1950 wrote that "The cost of wind turbines will no doubt decrease due to foreign research particularly in countries with critical shortages in fuel and hydro potential." But even when wind turbines become an economical alternative, they would remain a "contributor to the pool" of power sources, Whetstone said.[35]

Large-scale wind turbines reached a second critical stage of development in the early 1970s as a result of the Arab oil embargo. Numerous countries responded by initiating alternative energy programs. Wind programs focused on the megawatt-capacity turbines.[36] The U.S. government implemented a wind energy program of its own in 1973 under the jurisdiction of the National Science Foundation (shortly absorbed into the Energy Research and Development Administration and rolled into the newly formed Department of Energy in 1977). The foundation, in turn, handed the program over to the engineers at the National Aeronautics and Space Administration (NASA) Lewis Research Center in Cleveland, Ohio to carry out. According to a December 1973 NASA press release, "The wind is plentiful, inexhaustible, and non-polluting and could supply a significant fraction of our nation's energy needs in the future. What is needed is a determined, sustained effort to make wind derived energy available on demand at reasonable costs." NASA hyped the research of M.I.T. professor William Heronemus, a former nuclear submarine designer turned wind energy enthusiast, who estimated that about 350,000 wind turbines, each rated between 1,250 and 2,000 kilowatts, would be needed to equal the amount of electrical power used by the United States in 1969. NASA outlined its wind energy program objectives in late 1973 to include:

- Studies, construction, and testing of wind energy conversion systems (rotor, transmission, generator, tower, and yaw and pitch control mechanisms) without storage.
- Studies, construction, and testing of energy storage systems (compressed air, electrolysis of water to make hydrogen, pumped-hydro, large electrical networking).
- Conducting meteorological studies to estimate the wind-energy of the nation and to determine favorable regions and sites for wind driven energy systems.
- Studies and identification of suitable applications for demonstration tests.

NASA also presented its planned goals over the next five years:

- Identification of cost-effective wind energy conversion systems.
- Prototypes of wind conversion systems in operation.
- Proven wind conversion components and subsystems.
- Proven energy storage systems that are cost effective.
- Demonstration systems for selected applications, with storage, ready for testing.
- An accurate estimate of the nation's wind energy potential.
- Prospecting techniques for selecting sites on which to place wind conversion systems.
- The identification of regions having suitable winds.[37]

At the start of the program, NASA reviewed existing large-scale wind turbine designs, such as German engineer Ulrich Hütter's experimental 100 kilowatt machine and the restarted turbine at Gedser, Denmark. NASA sourced components manufactured by private sector engineering and aeronautical firms, such as Westinghouse Electric, Grumman Aerospace, Boeing and Lockheed, to build its experimental models. Ronald L. Thomas, project engineer for the NASA wind energy program, told *Business Week* in early 1974: "We're not out to prove that the windmill will work – that has been proven for hundreds of years. We are searching for practical, economical systems to use wind as an additional source of energy."[38]

NASA, which was assigned to initiate and develop the U.S. government's wind energy program in 1973, tested numerous blade types, but settled on a two-bladed design. *Courtesy of the NASA Glenn Research Center, Cleveland, Ohio.*

NASA completed construction of its first wind turbine at the Plum Brook Test Station near Sandusky, Ohio in 1975. The turbine, known as the Mod-0, was built on a four-legged 100-foot tall truss tower and included a two-bladed rotor with a diameter of 125 feet. NASA believed the two-bladed turbine was more economical to build. The turbine could reach its 100 kilowatt capacity in 18 mph winds. Wind direction was sensed by a vane on top of the gearbox and monitored by an automatic yaw control system.[39] The Mod-0 allowed NASA to test many aspects of wind turbine design and apply these improvements to its next generation test models. The Department of Energy decided to build four replicas of the first NASA wind turbine, called Mod-0A, to test power quality, start-up and shutdown, and safety. The four Mod-0As, rated at 200 kilowatts each, were installed in 1977 at Clayton, New Mexico; Block Island, Rhode Island; Culebra Island, Puerto Rico; and Oahu, Hawaii. These turbines were either located at remote sites or next to small utilities. The blades were constructed of either laminated wood epoxy or fiberglass.[40] These turbines suffered from routine breakdowns and cost overruns. The turbine at Culebra, for example, experienced frequent shutdowns due to an over-speed condition caused by loose belts in the mechanical controls, an invasion of fungus in the microprocessor circuits, and corrosion on the connections to the utility grid. However, NASA believed these turbines successfully demonstrated that wind power could be safely integrated with a utility's existing operations. The turbines also paved the way for the construction of even larger NASA test units.[41]

In 1979, NASA, with design and construction help from General Electric, installed its first experimental megawatt-scale turbine at Boone, North Carolina. The so-called Mod-1 turbine, dubbed in the national media as "the monster," consisted of a 150-foot tower with two 100-foot long blades and a power generating capacity of 2,000 kilowatts, enough electricity for about 500 homes at the time. The turbine cost a whopping $5.8 million to build.[42] The Mod-1 also alerted NASA officials to environmental problems associated with their early turbine designs, in particular noise and television signal interference. Neighbors of the Mod-1 turbine at Boone complained: "The monster rattles windows, bounces cups and saucers and creates an irritating swish-swish."[43] The government responded by shutting down the turbine at night, in the early mornings and on weekends, while engineers worked to correct the problems. NASA was already aware of the potential for television signal interference since it erected the Mod-0A on Block Island, and eventually installed a cable television system to resolve the problem.[44]

The three Mod-0A turbines (L to R: Clayton, New Mexico; Culebra Island, Puerto Rico; and Block Island, Rhode Island) in full operation. *Courtesy of the NASA Glenn Research Center, Cleveland, Ohio.*

By this time, NASA already had larger wind turbines on the drawing board. The agency offered Boeing a contract to build three 2.5-megawatt turbines at Goodnoe Hills near Goldendale, Washington to test the integration of a group of large wind turbines with a utility and to observe how they worked in close proximity. These turbines employed an upwind design to help reduce noise and eliminate wind interference from the towers. The Mod-2 also used a slender conical tower instead of the truss tower. These turbines suffered mechanical problems during the first two years. However, these setbacks weren't enough to stop NASA and Boeing from successfully carrying out their proof of concept. Boeing built two additional Mod-2 turbines, one for Pacific Gas and Electric Company at Solano County near San Francisco, California and another for the Department of Interior's Bureau of Reclamation for comparative testing at Medicine Bow, Wyoming.[45] The bureau had earlier contracted with United Technologies and Karlskronavarvet Company of Sweden to build a 2-megawatt turbine, known as WTS-4, to test if it could supplement wind power for water needed to power a hydroelectric system.[46]

NASA conducted smoke tests at the Lewis Plum Brook Test Station near Sandusky, Ohio in 1982. *Courtesy of the NASA Glenn Research Center, Cleveland, Ohio.*

Jointly developed NASA/Boeing 3 megawatt turbine raised at Oahu, Hawaii in July 1987. *Courtesy of the NASA Glenn Research Center, Cleveland, Ohio.*

NASA's Mod-3 and Mod-4 designs included large- and medium-scale turbines, but they never realized implementation. Instead, the agency promoted the next generation of high-output two-bladed turbines known as the Mod-5A and Mod-5B projects. These turbines were rated above 3 megawatts. General Electric withdrew from the Mod-5A project citing poor sales forecasts for large wind turbines. Boeing, on the other hand, decided to go through with the development of the Mod-5B turbine. The company completed construction of the Mod-5B turbine at Oahu on July 1, 1987. The unit consistently produced 3.2 megawatts of power in winds ranging from 13 to 17 mph. The turbine was sold in 1988 to the local utility and integrated with the power grid. According to NASA, the Mod-5B set a power record in March 1991, producing 1,256 megawatt-hours of electricity, a record by any single wind turbine.[47]

of the tower. The vortex would drive a flywheel and generator at the base of the tower. Yen believed his tower turbine could out-power the standard wind turbines by generating upwards of 100 to 1,000 megawatts.[55] In an interview with *Popular Science*, Yen considered power-generating units measuring up to 1,800 feet high and 600 feet wide on the New York City skyline.[56] In the 1980s, German aerospace company Messerschmidt-Balkow-Blohm (MBB) set out to build the giants of the wind energy industry. MBB first constructed the "Growian," based on Hütter's earlier designs, and shortly after that erected a single-blade 5 megawatt turbine on a 394-foot tower, called the "Growian II." The giant blade of the Growian II measured 238 feet long and weighed 26 tons. Directly opposite the blade was a stubby 35-ton counterweight. The blade was expected to turn 17 rpm and annually generate up to 17 million kilowatt hours, or enough electricity for about 350 homes. Both MBB's Growian turbine designs failed to meet operational expectations and were discontinued.[57]

In the late 1970s, inventor Terry Mehrkam built commercially available four-bladed wind turbines with mostly off-the-shelf components at well below costs of the government-funded test turbines. *Courtesy of Paul Gipe, Tehachapi, California*.

# Chapter Five
# California's Winds

California has long been a fertile ground for things new and improved, sometimes resulting in a big splash. Take the discovery of gold in the northern part of the state in 1848, which is remembered in history as the California Gold Rush. In a short period, tens of thousands of workers and investors flocked to California in search of their fortune. Like most economic booms, few individuals would win, many would lose. Soon the shine of the business flamed out. However there are some lasting benefits to each boom and bust cycle. In the case of the Gold Rush, a powerful and populace state emerged. San Francisco grew from a sleepy town into one of the Western world's biggest cities.

Fast forward 130 years and California found itself at the beginning of a new rush. This one involved the rapid production of commercial-scale renewable energy, namely through wind power. But just as California was not the first source of gold discovered in the United States, the state was also not the first to erect wind farms, or clusters of wind turbines operating to feed to an electric utility. That distinction belongs to a site at Crotched Mountain near Greenfield, New Hampshire. In 1979, U.S. Windpower erected 20 30-kilowatt turbines. The operation didn't last long due to severe technical problems, and America's first wind farm was terminated in 1980.[1] Another early wind farm was erected at Goodnoe Hills, Washington in 1981. The facility consisted of three Boeing-built Mod-2 Department of Energy-designed turbines. These turbines, each with 300-foot-long blades on 200-foot towers, were rated at 2.5 megawatts apiece and fed power into the grid operated by the Bonneville Power Administration.[2] This wind farm also suffered technical problems and was later dismantled. California offered early wind turbine makers like U.S. Windpower (later renamed Kenetech Windpower) abundant winds, and most importantly, the political and business climate to stimulate the development of the world's first large-scale wind farms. Federal and state tax credits covered 40 to 50 percent of the capital investment of turbines used for California's first wind farms. PURPA required the utilities to purchase the power at the "avoided cost" rate, which at the time was the equivalent the utility would pay to generate power from their most expensive fuels. Pacific Gas and Electric Company, for example, paid between 5.3 cents and 7 cents per kilowatt hour for wind-produced electricity.[3]

The three primary areas in California that became fertile grounds for wind farm developers in the early 1980s were Altamont Pass east of Livermore; San Gorgonio Pass near Palm Springs; and Tehachapi outside Bakersfield. Each location had its own valuable wind characteristics. In San Gorgonio, when the sun heats up the air, causing it to rise in the mid-morning and afternoon, cooler air from the Pacific Ocean funnels through the pass to fill the void. The average annual wind speed at San Gorgonio ranges between 13 mph and 25 mph depending on the location, with the highest winds occurring during the hot summer months.

Micon 108 kilowatt wind turbines installed at San Gorgonio Pass, near Palm Springs, California, in the mid-1980s. *Courtesy of Keith Higginbotham, Long Beach, California.*

A Micon wind turbine built in the early 1980s at Altamont Pass, California. *Courtesy of Paul Gipe, Tehachapi, California*.

Altamont Pass attracted three of the earliest wind farm builders. Between 1981 and 1983, U.S. Windpower built the first 100 turbines in Altamont Pass. These turbines had three-legged, 60-foot-tall steel towers with three-bladed rotor diameters of 56 feet. Fayette Manufacturing, another U.S. turbine builder at the time, set up 50 turbines on 40-foot slim tubular towers with blade diameters of 50 feet. Wind Master, the third company in Altamont Pass, ordered its 50 turbines from a builder in Belgium. These turbines had tapered cylindrical towers measuring 72 feet tall, with a 69-foot blade diameter. All three companies offered turbines rated at about 50 kilowatts each.[4] In the first half year of wind farming, both U.S. Windpower and Fayette experienced frequent shutdowns and problems with their blades, including stress and fatigue cracks. But the companies quickly made the repairs and generated attractive power outputs.[5] U.S. Windpower's turbines in Altamont, for instance, produced about 1.5 million kilowatt hours of electricity.[6]

There was no set pattern to development of California's first wind farms. Real estate investor Bill Adams bought land in the San Gorgonio Pass with the intent of selling it to the large wind farm developers. None of the developers wanted to buy the land and Adams went into the wind farm business himself. He sold wind turbines to doctors, lawyers and other investors. The investment tax credits allowed under the law at the time helped individuals offset their personal income tax.[7] By 1984, California's wind farms produced four times more electricity than they did in 1983, enough to power about 20,000 homes. By 1985, 11 firms were involved with wind farm development in the state.[8] Overseas turbine manufacturers actively penetrated the U.S. wind farm market during this period. By 1986, half of the state's turbines were made in Denmark. By 1990, California's wind farms contained more than 17,000 turbines, with ratings ranging from 20 kilowatts to 400 kilowatts. Together these turbines generated more than 3 million megawatt hours of electricity, enough to power a city of 300,000. California, at the time, also produced 90 percent of the world's wind-generated electricity.

Not everyone was hospitable to California's wind rush. Congressman Fortney "Pete" Stark opposed extending tax breaks to wind farms in the early 1980s. "These aren't wind farms. They're tax farms," he said.[9] Sonny Bono, musician-turned-mayor of Palm Springs, led a fight in the spring of 1989 to stop the construction of hundreds of additional wind turbines in nearby San Gorgonio Pass.[10] Conservative media outlets often poked fun at the state's wind farms. According to a 1994 *Forbes* article, Kenetech's then chief executive Gerald Alderson refused to be photographed with his company's turbines. "Why so shy?" the *Forbes* reporter asked. "Among many people the windmills have become a standing joke, and Alderson doesn't want his company to be thought funny."[11] By this time, Kenetech had erected 3,479 turbines in Altamont Pass. The company's turbines produced power at 7 to 10 cents per kilowatt hour, compared to 4 cents per kilowatt hour from conventional fuels. "Kenetech would be out of business were it not for tax breaks and federal and state mandates that have forced people to buy its products," the *Forbes* article declared. "The mandated business with Kenetech amounts to a hidden tax that helps raise PG&E's rates 50% above the national average."[12]

Another major setback to the California wind industry was the expiration of the federal tax credit program in 1985 and falling oil prices, making wind-generated power even more expensive. Turbine makers in the United States, such as Zond, Fayette and FloWind, were suddenly thrown into financial peril, eventually ending operations in bankruptcy. Kenetech, the largest player in the California market, succumbed in 1996. Aerospace and power generation giants, such as Boeing, General Electric, and Hamilton Standard, also abandoned the wind sector during this period. The effects in the U.S. market were felt overseas as well, particularly among the Danish turbine makers. Maintenance commitments from some manufacturers disappeared, further stoking the wind industry's image of poor reliability. U.S. wind industry historian Robert Righter noted an important upside to the industry's implosion in the late 1980s and early 1990s. "The bust was culling inferior machines and corrupt opportunists; leadership and integrity might win out," Righter wrote in his 1996 book *Wind Energy In America: A History*.[13] Indeed, the stronger players, particularly in Europe, survived and improved their turbines through better engineering.

The mid-1990s ushered in a movement to "re-power" California's less efficient wind turbines. Re-powering is the process of tearing down a site with old, unreliable turbines, reclaiming the land back to its natural state, and then installing a new project with newer, larger, more efficient turbines, while maintaining the existing substation and interconnection rights of the old project. The benefit of this activity to wind farms is increased annual production due to more efficient and higher capacity yielding turbines. In addition, the process restarts the 20-year life clock of the wind farm's turbines.[14] Proponents of re-powering believed this effort could also "improve public opinion" of wind turbines and help "restore California to its once prominent position as a showcase for renewable energy."[15] Re-powering in California's wind farms has taken off more so in recent years. Wintec Energy, which operates turbines on 1,200 acres in the Coachella Valley north of Palm Springs, announced in 2007 a $3 billion to $4 billion plan to upgrade the wind farm by tearing down the old machines and installing about 1,100 new turbines each separated by a quarter mile.[16] Similarly, Mesa Wind Farm, located on about 440 acres in the San Gorgonio Pass, plans to reduce its number of wind turbines from about 460 to 200 by installing new, larger units.[17]

A Kenetech 300 kilowatt turbine with a 108-foot (33-meter) rotor diameter installed during the mid- to late 1980s undergoes repowering work. *Courtesy of Paul Gipe, Tehachapi, California.*

California's leadership role in wind energy production dwindled in the late 1990s, quickly falling behind record development and output in Europe. It wasn't until the severe blackouts in the San Francisco Bay area in 2000 that many Californians seriously renewed their interest in wind power.[18] But it's perhaps too late to restore California to its once dominant spot in the world of wind energy. Even in the United States, the state's reign appears to be over. Wind turbine installations are more expensive in California than most other states, due to permitting, various government regulations, and in general the cost of doing business. California wind farm developers continue to rely on higher power purchase prices to remain profitable. Texas already edged out California in 2006 with the most installed wind capacity. According to the American Wind Energy Association, Texas had cumulatively installed 2,768 megawatts by the end of 2006, compared to California at 2,361 megawatts. Also, by 2007, Texas hosted the country's largest wind farm at Horse Hollow with 736 megawatts.[19] With sufficient land and pro-wind state policies, Texas is on target to maintain this lead for the foreseeable future. In 2007, more than 3,000 megawatts of wind power capacity was installed Texas.[20]

Six General Electric 1.5 megawatt wind turbines standing at Karen Avenue in the San Gorgonio Pass near Palm Springs, California. The three units in the background were installed in 2003, while the three units in the foreground were built in late 2004. *Courtesy of Keith Higginbotham, Long Beach, California.*

California remains bullish on wind energy production to meet its environmental goals. The state aims to meet 20 percent of its electricity requirements through renewable energy sources, such as wind, by 2010. New wind farms have been proposed. On December 21, 2006, Alta Wind Power Development and Southern California Edison signed an historic energy power purchase agreement to generate 1,500 megawatts. Alta Wind is a partnership of Australia-based Allco Infrastructure and Oak Creek Energy in Tehachapi. The Alta Wind project, which is pending state approval in mid-2007, would cost about $3 billion to build and include 750 turbines across 50 square miles.[21] Western Wind is also reviewing more than 60 locations in California for a new wind farm site.[22] Portland, Oregon-based PPM Energy in 2007 started construction of 45 turbines in the Palm Springs area, which will generate power for about 21,000 average California homes.[23] There are applications for new sites, such as in Needles, California. Smaller wind farm sites already in existence, like those near Los Banos and Pacheco Pass, are also prepared for growth.

These General Electric 1.5 megawatt turbines were installed at San Gorgonio Pass near Palm Springs, California by Whitewater Energy Corporation in 2002 and sold to Shell Wind Energy. *Courtesy of Keith Higginbotham, Long Beach, California.*

Portion of the 120 megawatt Hadyard Hill wind farm in South Ayrshire, Scotland in the United Kingdom. Commissioned in March 2006 for Scottish and Southern Energy, the project consists of 52 Siemens turbines, each capable of generating 2.3 megawatts of power. By 2006, this was the most powerful wind farm operating in the United Kingdom. *Courtesy of the British Wind Energy Association*.

By 2006, Spain – home of the legendary tale of Don Quixote, the old man who tilted at windmills – became Europe's second largest producer of wind energy at 11,615 megawatts.[15] Spain's wind turbine installations steadily increased after the government passed special legislation in 1997 granting renewable energy production guaranteed access to the grid and premium payment for generated power. By the end of 1997, Spain's wind farms generated upwards of 200 megawatts, followed by annual growth rates on average of 30 percent a year through 2003. The country installed 8,263 megawatts of wind power by the end of 2004, a 6.5 percent contribution to the national electricity supply.[16] The first installations were built in the southern province of Andalucia. These were followed by wind farms in the northern regions of Galicia and Aragon, continuing onto the central provinces of Castilla-La Mancha and Castilla y Leon. By 2006, Galicia led all the regions with 2,603 megawattts, closely followed by Castilla-La Mancha at 2,311 megawatts, and Castilla y Leon at 2,120 megawatts. Turbines are found throughout all of Spain, including the Canary Islands and Baleares in the Mediterranean, meeting 9 percent of the country's electric power requirement by the end of 2006.[17] The country boasts more than 500 companies involved in wind energy, with most of its turbines being made domestically.[18] The biggest companies involved in Spain's wind farm development are EHN of the Acciona Group, Corporacion Eolica SA, and Iberdrola, which became the largest renewable-energy utility by 2006. Gamesa has become the country's largest world recognized turbine manufacturer. Spain remains bullish on wind power and plans to install at total of 20,155 megawatts of power by 2010.[19]

The El Perdon wind farm in Navarra, Spain. Consisting of 37 Gamesa 500 kilowatt and three Gamesa 600 kilowatt turbines, the project was constructed between 1994 and 1996. *Courtesy of Corporación Energia Hidroeléctrica de Navarra, Navarra, Spain/European Wind Energy Association, Brussels, Belgium.*

More than 100 Gamesa turbines of the La Plana IV wind farm dot the landscape of the Aragón region of Spain. *Courtesy of Vestas, Randers, Denmark.*

Other European countries, which have carved out substantial stakes in wind energy by 2006, were Italy (2,123 megawatts), Portugal (1,716 megawatts), France (1,516 megawatts), and the Netherlands (1,560 megawatts).[20] Italy's wind energy program took off in 1998, with most of its turbines operating in the southern part of the country. Italy's wind turbines range in size from 500 kilowatts to 2 megawatts, but the larger the size the more difficulties are faced, particularly with the hilly terrain and limited transportation access to certain ideal sites.[21] Portugal faces similar limitations due to its terrain, park land restrictions, and less than adequate transmission lines.[22] The country installed 1,055 turbines by 2006, meeting 59 percent of its 2010 wind energy target.[23] By the end of 2006, 48,062 megawatts of wind projects were installed throughout the European Union, which could produce on average 100 terawatts of electricity a year, or 3.3 percent of the EU electricity consumption in 2005.[24] In 2007, one of the technical drawbacks within the European Union is the inability to efficiently interconnect wind-generated electricity with the various national power grids. "For large amounts of wind power to be integrated into the grid, closer cooperation is needed between the wind industry and utilities, systems analysts, grid operators, regulators and decision makers," wrote Christian Kjaer, chief executive of the European Wind Energy Association (EWEA), for *EU Power*. "Significant investments in the transmission network are crucial to improving the functioning of the European internal electricity market. Electricity networks of the future will have to be better connected and reshaped."[25]

A 2 megawatt G80-2000 Gamesa turbine at Florinas on the northern coast of Sardinia, Italy. *Courtesy of Gamesa Corporacion Tecnologia SA, Madrid, Spain/European Wind Energy Association, Brussels, Belgium.*

Three of the 25 General Electric turbines used at the Haute-Lys wind farm in Pas-de-Calais, France. *Courtesy of Syndicat des Energies Renouvelables – France Energie Eolienne, Paris, France*.

A French high-speed TGV Réseau train sails past several Vestas V82 1.65 megawatt turbines in the Bois-de-Bigot wind farm southeast of Paris, France. *Courtesy of Syndicat des Energies Renouvelables – France Energie Eolienne, Paris, France.*

A Vestas V90 3 megawatt turbine being erected for the energy firm Raedthuys in the Netherlands. *Courtesy of the Nederlandse Wind Energie Associatie, Ultrecht, Netherlands.*

Eastern Europe is considered the new frontier for the continent's wind farm developers. According to Emerging Energy Research, Eastern Europe's wind power capacity is anticipated to grow from 569 megawatts in 2006 to 7,552 megawatts by 2015. Most of this expansion has been witnessed in Poland, Lithuania, and Hungary.[26] Some of Europe's largest wind energy firms have moved into the market, such as Spain's Acciona and Iberdrola, EuroTrust of Denmark, and Good Energies in the United Kingdom. Financial support for wind energy projects in Eastern Europe has also come from the European Bank for Reconstruction and Development.[27] Poland is considered to have the best potential for widespread wind energy use in Eastern Europe due to friendly government policies toward renewables. The country installed 153 megawatts of wind capacity by 2006, and increased that amount to 175 megawatts at the start of 2007 with the commissioning of the Wind Farm Puck.[28] As a condition for Poland's accession to the European Union, the country must produce 7.5 percent of its energy from renewable energy sources by 2010. The Polish government plans to build another 1,800 megawatts of wind projects between 2007 and 2010, or about 450 megawatts a year.[29] In May 2007, Good Energies said it will invest $234 million to build a 120-megawatt wind farm – enough to power 72,000 homes – in Poland by late 2008, and the company has additional plans to erect upwards of 500 megawatts of installed wind energy capacity in the country by 2010.[30] The Global Wind Energy Council warned that meeting a 2,000 megawatt threshold through wind energy by 2010 will be difficult for Poland to attain because of insufficient financial supports for investment, lack of project evaluation measures and rules to apply funds, problems with connections to the power grid, and environmental restrictions. "The Polish wind energy market is in its early stages of development, but wind conditions are comparable to countries currently leading (the) sector, and the overall potential of Poland is large," the Global Wind Energy Council said. "With favorable conditions and consequent elimination of barriers to wind energy development, the perspectives for the sector are quite positive."[31]

Vestas V80 2 megawatt turbines at the Energia Eco Ltd. Cisowo wind farm on the northern coast of Poland. *Courtesy of Vestas, Randers, Denmark.*

REpower Systems AG MM-82 2 megawatt turbines overlooking a riverside Nike distribution center at Laakdal, Belgium. *Courtesy of Peter Neyens, Sint-Truiden, Belgium.*

In addition to Europe's political will, many large businesses on the continent actively support the purchase of renewable energy and some have even deployed wind turbines with the specific intent of powering their factories and warehouses, and to demonstrate their environmentally friendly practices to the public. Footwear manufacturer Nike, for example, uses six wind turbines with a total output of 9 megawatts at its product distribution center in Laakdal, Belgium. The turbines are owned and operated by German firm SeeBA Energy Systems and fully meet the power needs of the Nike facility. The turbines also helped to fulfill Nike's promise to the World Wildlife Fund to invest in green energy, called the Climate Savers Program. In addition, Nike as a company reduced its emission of greenhouse gases worldwide by 13 percent in 2005, and has continued to invest in energy saving air conditioning systems, lighting and other equipment.[32]

In 2006, the European market for wind turbines had grown 23 percent, compared to the previous year. This included the installation of 7,588 megawatts of capacity, valued at about $12 billion. The cumulative wind power capacity operating across the European Union's 27 member states exceeded 48,000 megawatts. According to the EWEA, in an average year this capacity will generate about 100 terawatts of electricity, equal to 3.3 percent of the European Union's total energy consumption.[33] In 2003, the EWEA set a 2010 wind energy capacity goal of 75,000 megawatts among the initial EU-15 member states, but with the additional 10 member states which joined the union in 2004, the collective output could reach as high as 80,000 megawatts by 2010.[34] Overall by 2007, the European renewable energy industry had a turnover of 25 billion euro and a workforce of 300,000, making it one of the fastest growing sectors in the economy.[35]

In March 2007, the leaders of 27 European Union states reached a decision to adopt a binding 20 percent target for renewables by 2020 with the realization that wind energy will play the biggest role, possibly up to 16 percent.[36] The European Union hopes the use of renewables will make a dent in harmful emissions and give it leverage when dealing with oil-producing countries. "If we want to be in a strong negotiating position with countries that have a rich supply of oil and gas, we need to be able to prove that we are able to produce energy from local resources," Andris Piebalgs, European Commissioner for Energy, told EWEA's *Wind Directions* magazine in mid-2007. "This is the only way to convince these countries to be reasonable in their demands and sell their commodity at a fair price – not to make exorbitant demands."[37] The European Union expects its energy imports to increase from 50 percent in 2007 to 65 percent by 2030.[38] With stable policies in place, investments in wind technology development should increase. The EWEA forecasts wind energy output in Europe should reach up to 300 gigawatts, enough to power 200 million households or satisfy 22 percent of the continent's energy need, by 2030. "If the new European target is achieved, more than a third of Europe's power will be renewably sourced by the end of the next decade (2020)," said Peter Ahmels, president of the German Wind Energy Association, in a July 27, 2007 *EnergyBiz Insider* article. "This has sent a clear political signal. It is no less than the beginning of a second industrial revolution."[39] Ahmels estimated that millions of new jobs would be created through Europe's renewable energy targets.[40]

To illustrate pride in its wind energy acceptance, the EU celebrated its first European Wind Day on June 15, 2007. The event was marked by special activities across the continent, including wind farm tours, competitions, school lectures, press conferences, and exhibitions. The EWEA declared "European Wind Day will be the perfect occasion to learn about wind energy, have fun and support clean electricity. Together, we can celebrate the power of wind!"[41]

Wind turbine blades and towers being positioned at Intermarine's Industrial Terminals in Houston for later delivery to U.S. wind farms. *Courtesy of Intermarine, New Orleans, Louisiana*.

To maintain and grow this business, ports must invest in equipment to efficiently handle larger volumes of turbine components. The Port of Vancouver USA, for instance, acquired the largest harbor crane in North America in 2006 to assist with its burgeoning wind turbine business. The $3.2 million 500-ton LHM 500S crane, built by Austrian manufacturer Liebherr, is capable of hoisting 140 metric tons with up to 60 feet of outreach. With 80 wheels on 20-axle sets, the crane can turn on a dime in any direction, which helps minimize additional handling of turbine components.[7] Bulk ports like Vancouver also offer sufficient space for storing turbine parts until they are ready for transport to the wind farm sites. In 2006, Longview added nine acres of storage yard and purchased new Hyster forklifts. A Kalmar reach stacker, purchased by the port in 2005, is used extensively to move tower and turbine pieces.[8]

Port equipment prepares a blade for truck transportation. *Courtesy of Port of Vancouver USA, Vancouver, Washington*.

Some U.S. ports have invested in special heavy lift equipment for handling wind turbine component shipments. In 2006, the Port of Vancouver USA in Washington (pictured above) acquired a 500-ton LHM 500S crane from Liebherr to hoist increasingly larger turbine pieces. *Courtesy of the Port of Vancouver USA, Vancouver, Washington*.

Before the turbine components can leave either the domestic manufacturing site or port, the proper oversize and overweight permits must be obtained from the state authorities. Each turbine generally requires 10 truckloads, eight of which need special transportation permits.[9] Components must be loaded on the right types of equipment. Dallas, Texas-based LoneStar Transportation, one of the largest domestic transporters of turbine components in the United States, developed a fleet of specialized multi-axle lowboy blade trailers. LoneStar continues to invest in the development of new trailers to transport increasingly larger turbine components.[10] According to the freight transportation publication *American Shipper*, transportation costs per individual turbine may range between 2 percent and 20 percent.[11]

The next step in the logistics process is to deliver the turbine components to the wind farm sites. Although some parts are moved by railcar, the majority are transported by truck. Transportation routes must be carefully studied for potential obstacles. Stephen Donchez, president of Vineland, New Jersey-based American Transport Systems, is a cartographer by training and often visits the wind farm sites to observe first hand the condition of the roads and bridges. "You literally work back from the X on the ground," Donchez told *American Shipper* in a September 2003 article about managing the transportation of wind turbine components.[12] David Ferebee, vice president of sales and marketing for LoneStar, emphasized the importance of having managers on site at both the port and wind farm throughout the transportation process. "Transportation for the project then becomes a seamless issue," he said.[13] Wind farms in the U.S. Northeast are generally more difficult to access because of narrower roads, older bridges, and community density. Even in the Western states, once trucks leave the main roads for the job site, the transportation can quickly become difficult. Inclement weather may halt the transportation altogether.

Heavy duty truck loaded with nacelle begins journey to wind farm site.
*Courtesy of Port of Vancouver USA, Vancouver, Washington.*

Once the concrete pads are poured and ready, special construction crews with heavy lift cranes are brought in to lift the giant turbine components into place. The tower is erected first. Next, the nacelle, which contains the rotor and generator equipment, is set on the top of the tower. This is followed by the attachment of the blades to the rotor. Lastly, microprocessors are installed to control the blade movement and regulate the output to the underground power lines. All this equipment must meet a series of building codes and pass a stringent engineering review. These final checks help to ensure that the turbines can withstand catastrophic events, such as high winds, earthquakes and floods. Occasionally total losses do occur, but are often due to mechanical malfunctions rather than natural forces.

Phases of construction for a series of three General Electric 1.5 megawatt wind turbines along Karen Avenue in the San Gorgonio Pass near Palm Springs, California, in September 2004. The process starts with the base, and then installation of towers and nacelles. This is followed by staging of the blades and cranes to lift assembled blade rotors into place. *Courtesy of Whitewater Energy Corp., Torrance, California.*

Wind turbines are spaced in either rows or grid patterns to best take advantage of the wind. *Courtesy of Keith Higginbotham, Long Beach, California*

Depending on the location and wind characteristics of the site, turbines are generally spaced in two formats. The upwind/downwind layout requires a space between turbines of 5 to 10 rotor diameters. This is the distance of how far back the downwind turbine should be set to receive smooth airflow or no turbulence from the upwind turbine. In California's San Gorgonio Pass, 70 percent of the wind enters from the West, 20 percent from the East, and 10 percent from other directions. Therefore, due to the predominant West/East direction of the winds, and very little wind from the North or South, turbines are generally spaced close together. Older turbines in the pass are spaced about 50 feet apart from blade tip to blade tip. For new turbine installations in San Gorgonio, manufacturers require at least a 1.25 rotor diameter space from the center of the tower to the blade tip of the neighboring turbine. The other format for turbine spacing is the grid. This formation is often found in offshore wind farms or those facilities located on large expanses of open plains. These sites generally receive equal amounts of wind from all directions. Each turbine would therefore be spaced five to 10 rotor diameters from each other both side to side and upwind/downwind to form a grid pattern.[14]

Compared to other power generation projects, wind farms take the least amount of time to erect – often between three and six months. However, since 2005, U.S. wind farm developers have experienced a tight supply market and rising materials costs. Industry analysts blame supply constraints on several factors, including the renewal of the federal government's wind energy tax credit program in late 2004, rapidly approaching renewable energy state usage deadlines, increased order sizes both in volume and turbine capacity, rising international competition for wind turbine components, and steadily higher prices for base production materials, such as steel, copper and carbon. Added up, wind turbine parts manufacturers find it difficult to meet their orders. Some of the tightest points in the turbine component supply chain are associated with the blades, gearboxes, bearings and other cast iron and forged pieces. Two leading wind market analysts, BTM Consult and Emerging Energy Research, independently concluded that the supply chain constraints for turbine suppliers would not stabilize until 2009.[15] The 1.5 to 2.5 megawatt turbines in 2007 reached costs of $3 million or more per unit.

Since 2005, some large overseas turbine manufacturers and related component suppliers have set up production facilities in the United States to help cope with rising demand and to shorten their delivery cycles to wind farm sites. For example, Spanish turbine manufacturer Gamesa took over 20 acres at a former steel plant in Ebensburg, Pennsylvania to build the blades, nacelles and towers for its 2 megawatt turbines. Suzlon Wind Energy of India set up a blade manufacturing and tower assembly facility in Pipestone, Minnesota, and Mitsubishi Power Systems of Japan now builds gearboxes at a plant in Lake May, Florida.[16] Denmark's Vestas constructed a blade and turbine manufacturing plant on 100 acres of industrial property near Windsor, Colorado, in an effort to more efficiently supply wind farm developers in the Rocky Mountain region. These developments bring with them the promise of hundreds of skilled jobs and revitalization to so-called "rust belt" areas of the country.[17] The wind farm construction frenzy in the United States also continues to strain other industries with ties to building activities such as providers of cranes and specialized transportation equipment.[18]

Last and most important, operating an efficient wind farm requires regular maintenance. The general maintenance schedule includes annual and semi-annual services on each turbine. There are also the unscheduled maintenance calls for breakages as turbines become older. Many wind farm owners will "retrofit" poorly designed, mechanically troublesome, and inefficient turbines, which may include the replacement of gearboxes with different ratios to allow the turbine to spin a little faster to produce more energy in lower winds, or change from a high-speed to a low speed brake system to reduce the strain on the gearbox and blades due to the high braking force.[19] (With many early design flaws engineered out of today's turbines, retrofitting is quickly becoming a thing of the past.) Blades will experience edge wear and occasional surface cracking due to wind erosion and changing temperatures and moisture content in the air. Lightning strikes may also inflict severe damage to blades. Companies that specialize in these types of repairs must be contracted to do the work.[20] Over time, it becomes increasingly difficult for maintenance companies to acquire parts from original manufacturers to service older turbines. Thus they must reach out to the second-hand parts market or machine shops for customized pieces. Whitewater Maintenance Corporation in the San Gorgonio Pass, for instance, regularly orders custom made parts, such as main shafts, gears for gearboxes and yaw systems for its older turbines. When electronic components are discontinued, Whitewater either hires an electrician to rebuild the circuit boards or asks the electronics company to back-engineer the part and provide it with new ones.[21]

Today's wind turbines are rugged and with routine maintenance can operate continuously for many years. Occasionally, there are severe malfunctions. This turbine (picture above) caught fire on December 23, 2005 at a Nissan factory in the United Kingdom. The fire was reportedly visible for miles across the Wearside area. *Courtesy of the Northeast Press/ Sunderland Echo, Sunderland, United Kingdom.*

The increased wind farm activity throughout the United States keeps maintenance firms scrambling for skilled technicians. Some community colleges and technical schools have started offering courses tailored to wind turbine maintenance. According to the Columbia Gorge Community College in Dalles, Oregon, wind farm development in both Oregon and Washington through 2015 will require about 365 new technicians. The number of technicians increases at a rate of about eight people per 100 megawatts of installed power.[22] The country's largest wind farm developer, FPL Energy, entered a partnership with the Texas State Technical College in 2007 to serve as an advisor to the school's two-year degree program for wind turbine maintenance. "As the largest owner and operator of wind turbines in Texas and throughout the world, having a well trained workforce is critically important to our future growth and ultimate success," said Manny Sanchez, vice president of wind operations at FPL Energy, in a Feb. 15, 2007 statement.[23]

However, turbine maintenance isn't suited for just anyone with mechanical aptitude. Technicians must be able to endure heights of up to 250 feet and be willing to work in cramped quarters for up to 12

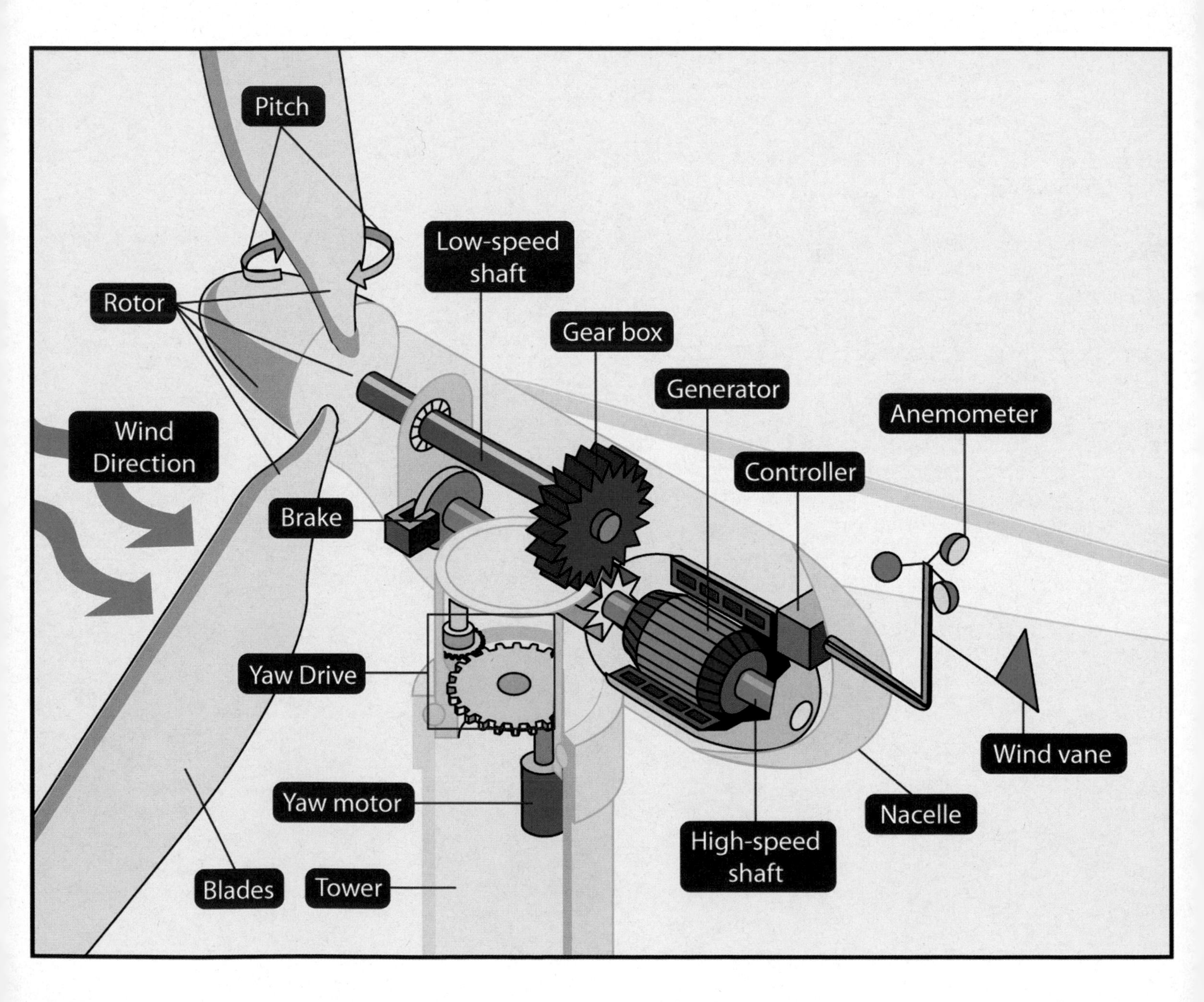

hours at a time. The job is also dangerous. Turbines have many large moving parts, which can instantly maim or kill.[24] American wind energy expert Paul Gipe on his Website maintains a worldwide tally of wind turbine-related deaths dating back to 1975. Its purpose, he writes, is "so that everyone who works with wind energy, or who contemplates working with wind energy, will carefully weigh their actions."[25]

**Basic wind turbine parts**

Anemometer – Measures the wind speed and transmits the data to the controller.

Blades – Most turbines have either two or three blades. Wind blowing over the blades causes them to lift and rotate.

Brake – A disc brake, which can be applied mechanically, electrically, or hydraulically to stop the rotor in emergencies.

Controller – The controller starts up the machine at wind speeds ranging from 8 to 16 mph and shuts off the machine when winds reach about 65 mph. Turbines cannot operate at wind speeds above 65 mph because their generators could overheat.

Gearbox – Gears connect the low-speed shaft to the high-speed shaft and increase the rotational speeds from about 30 to 60 rpm to about 1,200 to 1,500 rpm, the rotational speed required by most generators to produce electricity. Some turbine builders are considering "direct drive" generators that operate at lower rotational speeds and don't need gearboxes.

Generator – An induction generator that produces 60-cycle AC electricity.

High-speed shaft – Drives the generator.

Low-speed shaft – The rotor turns the low speed shaft at about 30 to 60 rpm.

Nacelle – The rotor attaches to the nacelle, which sits atop the tower and includes the gearbox, low- and high-speed shafts, generator, controller, and brake. A cover protects the components inside the nacelle. Some nacelles are large enough for a technician to stand inside while working.

Pitch – Blades are turned, or pitched, out of the wind to keep the rotor from turning in winds that are too high or too low to produce electricity.

Rotor – The blades and the hub together are called the rotor.

Tower – Towers are made from either tubular steel (shown here) or steel lattice. Since wind speed increases with height, taller towers enable turbines to capture more energy and produce more electricity.

Wind direction – Pictured is an "upwind" turbine, because it operates facing into the wind. Other turbines are designed to run "downwind," or facing away from the wind.

Wind vane – Measures wind direction and communicates with the yaw drive to orient the turbine properly with respect to the wind.

Yaw drive – The yaw drive is used to keep the upwind turbine facing into the wind as the wind direction changes. Downwind turbines do not require a yaw drive, because the wind blows the rotor downwind.

Yaw motor – Powers the yaw drive.

For additional wind energy terms, refer to the glossary in the appendix.

*Courtesy of the National Renewable Energy Laboratory, Golden, Colorado.*

Brown, which is also developing onshore wind farms in New Mexico, South Dakota, California, and Hawaii, stepped away from the deal in May 2007, citing that the project was not economically feasible. The Texas Land Commission, although disappointed by Babcock & Brown's decision, believes another firm will be found.[17] Texas has one of the most unusual coastal regulatory authorities in the United States. While other states have rights over submerged lands up to 3 nautical miles off their coasts, Texas' authority stretches 9 nautical miles, or three nautical leagues, from its coastline. This difference dates back to 1836 when the Republic of Texas won independence from Mexico. Texas entered the Union in 1845 with its offshore boundaries intact. Today, all offshore wind farm applications run through one state office – the Texas Land Commission – for approval, not multiple offices as is the case in most states.

Denmark's experience in offshore wind farms commemorated by Post Denmark in a postage stamp series released on January 10, 2007.

The strong winds of the Great Lakes also offer the opportunity for offshore wind farm development. A 2004 report funded by the Wisconsin Focus Energy Program found promising wind activity along the state's southeast coast with Lake Michigan. Steady winds of 19 mph have been clocked in the area within a few miles from shore. "Should (turbine blade) hub heights above 80 meters (262 feet) be used, wind resources would be even higher," the report concluded.[18]

Offshore wind farms offer an opportunity to erect larger turbines than what are found on land. Turbines currently used in most offshore wind farms have towers upwards of 260 feet tall, with individual blades measuring about 130 feet. They each produce between 1.5 megawatts and 3 megawatts of power, but larger outputs are planned. Because of the larger turbine size and harsh environment, offshore wind farms pose an entirely different set of construction challenges than landside projects. First, offshore turbines must have strengthened towers to withstand ocean waves. Nacelles must be properly shielded from the salty air to protect the internal generator equipment from corrosion and capped with navigational and aerial warning lights. To minimize onsite maintenance, these turbines require efficient self-lubricating systems. There are also the logistics involved with connecting the offshore wind power to land. Each turbine has its own transformer to bring the voltage to a distribution level of about 34,000 volts. If the project is big enough, there will be an electric service platform at sea where the voltage is increased to levels four to eight times higher and brought ashore. The interconnection on shore will have its own requirements. Cables are buried under the sea floor in all cases.

Initial lifting of monopiles at Kentish Flats offshore wind farm in 2005. *Courtesy of MPI Offshore Ltd., Middlesbrough, United Kingdom.*

Before starting construction, a turbine's components are generally shipped from various factories to a marine terminal nearest to the offshore site where large sections are pre-assembled, such as the towers, blade units and nacelles. In Northern Europe, there are a handful of companies that focus on the transport and setup of offshore wind turbines, such as A2SEA of Denmark, SeaRoc and Marine Projects International (MPI), both based in the United Kingdom, and Dredging International of Belgium. These firms operate small fleets of specialized vessels. For example, A2SEA has two vessels, each equipped with jacking legs to raise their hulls in the water to provide a stable platform on which to operate. On the deck of each vessel is a 450-ton crane with 110 tons of lift, capable of raising turbine components about 260 feet high and up to 70 feet away from the side of the ship. There is a smaller crane on the bow of each vessel to handle job containers and smaller components. MPI operates the *Resolution*, currently the largest self-elevating vessel in the offshore wind farm installation business. In addition to cranes for setting up the turbines, the *Resolution* includes a 160-ton onboard pile hammer and cable laying equipment.[19] Traditionally, these vessels transport components for three to four turbines per voyage to the offshore site where another company has preset the platforms on which the turbines will stand. These platforms are either concrete gravity foundations, which rest on the sea bottom, or steel monopiles, which must be drilled or vibrated about 65 feet into the seabed. These foundations sit in water depths of less than 30 meters. The winds that keep these large turbines turning are the same that determine the schedule for installation. A2SEA estimated that in ideal weather it can install a turbine in 12 hours, but the average installation takes about one and a half days. Nearly 70 percent of the installation work for offshore wind farms takes place during the summer. The work must stop, however, when the winds reach over 25 mph and waves exceed six feet high.[20]

Some experts have suggested combining wind power with other forms of offshore energy production, such as natural gas and tidal systems. In February 2007, the United Kingdom approved a "hybrid" offshore wind farm to be located about 10 kilometers off Walney Island near Barrow. The "Ormonde project" of Eclipse Energy will combine wind turbines with natural gas power. The U.K. Department of Business, Enterprise and Regulatory Reform said the project has the potential to generate up to 200 megawatts of electricity, with about half coming from a 30-turbine wind farm. The concept is that when the wind is insufficient for the turbines, power will come through conventional gas sources pumped from the offshore fields in nearby Morecombe Bay.[21] The United Kingdom is also an advocate of tidal power, which involves the installation of underwater machines that look quite similar in design to wind turbines.

No doubt offshore wind power will evolve. Dr. Eddie O'Connor, founder and former chief of Airtricity in Ireland, recently proposed to develop a "Supergrid," that would provide a series of interconnections for offshore wind farms throughout the waters surrounding Europe. To launch the Supergrid concept, Airtricity would implement its "10 GW Foundation Project" in the North Sea, interconnecting its offshore wind farm output to power grids in the United Kingdom, Holland and Germany. "It would be the largest offshore wind park in the world, with each turbine capable of generating 5 MW (megawatts) of electricity. The turbines would be rooted to the sea floor with the transmission running along the seabed. The project would link the markets of the three countries involved," O'Connor wrote in an article for trade journal *Windtech International*.[22] Airtricity proposed to begin construction of the 10 GW Foundation Project in 2011 with a completion date of 2017. In theory the project would consist of three offshore sites, one off the United Kingdom's Thames Estuary, and two other sites, yet to be determined, off the coasts of Germany and Holland. O'Connor warned, however, that the biggest obstacle to the Supergrid would be gaining acceptance of the concept from the European Union and national government levels.[23]

A monopile upending tool guidance frame. *Courtesy of MPI Offshore Ltd., Middlesbrough, United Kingdom.*

Loading a turbine tower for the Netherlands' Egmond ann Zee offshore wind project in the summer of 2006. *Courtesy of A2SEA, Fredericia, Denmark.*

Wind turbine blades assembled in the fashion, known in the industry as "bunny ears," on the terminal before loading on vessels. These bunny ears and blades are headed for the North Hoyle offshore wind farm off the United Kingdom's Wales Coast in 2003. *Courtesy of MPI Offshore Ltd., Middlesbrough, United Kingdom.*

Transportation of bunny ear assemblies and third blades for North Hoyle offshore wind farm in 2003. *Courtesy of MPI Offshore Ltd., Middlesbrough, United Kingdom.*

The offshore wind turbine installation vessel *Resolution* jacked up for stability next to the North Hoyle offshore wind farm in 2003. *Courtesy of MPI Offshore Ltd., Middlesbrough, United Kingdom.*

Installation of bunny ear assembly at the North Hoyle offshore wind farm in 2003. *Courtesy of MPI Offshore Ltd., Middlesbrough, United Kingdom.*

Installation of a third blade to complete the rotor and blade assembly of a turbine at the North Hoyle offshore wind farm in 2003. *Courtesy of MPI Offshore Ltd., Middlesbrough, United Kingdom.*

Installation of array cables with MPI Offshore's ROV. *Courtesy of MPI Offshore Ltd., Middlesbrough, United Kingdom.*

Proposed deep sea platforms in 2006 for offshore wind turbines. *Courtesy of the National Renewable Energy Laboratory, Golden, Colorado.*

Offshore wind turbine installation vessel finishes a day's work at the Horns Rev offshore wind farm off the Danish coast in the autumn of 2004. *Courtesy of A2SEA, Fredericia, Denmark.*

In addition, there is increased discussion within the scientific and engineering communities about erecting turbines in much deeper waters. In 2006, Paul D. Sclavounos, a professor of mechanical engineering and naval architecture at the Massachusetts Institute of Technology, suggested that someday more than 400 large wind turbines could be set up about 100 miles at sea and feed electricity landside at a capacity sufficient to power more than 300,000 homes.[24] M. I. T. worked with the Department of Energy's National Renewable Energy Laboratory on the concept of floating individual wind turbines in the capacity range of 5 megawatts. Since these extra large turbines would be cost-prohibitive to set up at sea, Sclavounos proposed to build them onshore in a shipyard and use tugboats to tow them to the sites. To keep the platforms stable, they would be ballasted with water and concrete. Once the tethers of the floating platforms are secured to the seabed, the ballast would be removed until the tethers are tight. According to Sclavounos and his team, the tethers would allow the platform to move side by side but not up and down, as vertical motion could hinder the turbine's function. The turbine platforms would have a diameter of 30 meters. The turbine blades would be high enough on the tower to avoid dipping into the biggest waves.[25] Because the turbines would be farther out to sea, they should be able to generate more power than the existing offshore turbines. There would also be the added benefit of being able to unhook the tethers and float the turbines to other sites.[26]

# Chapter Nine
# North America's Wind Rush

From the late 1980s to the mid-1990s, the United States paid little attention to renewable energy. Attempts both from within the government and private sector to fuel the wind industry were feeble at best. The Carter administration in the late 1970s introduced a subsidy, known as the Investment Tax Credit (ITC), to stimulate construction of commercial wind turbines, but that incentive died as part of the 1986 Tax Reform Act, and the United States lost most of its nascent wind turbine manufacturing capability within a few years of its expiration. Petroleum-based fuel prices also fell during the 1980s, adding to the deteriorating public interest in renewable energy sources. Industry giants such as Boeing and General Electric abandoned the wind business in the mid-1980s shortly after the ITC lapsed, resulting in the layoff of many aerospace engineers. Some of these engineers, however, didn't give up hope and set out to develop their own wind energy companies.[1]

A new incentive was put in place in 1992 – the wind Production Tax Credit (PTC). The PTC is essentially a federal rebate on the taxes paid by companies that own wind projects. Congress, which is responsible for reauthorizing the PTC, allowed the benefit to periodically lapse, including a disastrous eight month lapse in 2004. When Congress did succeed in reestablishing the PTC, it was not for more than a year or two at a time, causing the wind energy sector to go through "on-again, off again" production cycles. When the congressionally mandated expiration date for the PTC got closer, wind energy developers rushed to get their turbines in the ground. A wind project could not enjoy the benefits of the PTC unless it was brought online before the expiration.

The public attitude toward wind energy began to change in the mid-1990s as fuel prices began to climb and heightened outcry from the scientific community that pollution from oil and coal burning power plants may be exacerbating global warming. With the PTC in place and periodically extended, and with the cost of wind starting to decline very steeply as the technology matured, the wind energy business finally started to move forward because of policy leadership in states such as Iowa and Minnesota. The PTC provides a 2.0 cents-per-kilowatt-hour tax credit for electricity generated by wind turbines over the first 10 years of a project's operations.

Another positive development for the wind sector in the late 1990s was the introduction of many state-based Renewable Portfolio Standards (RPS) programs, which set specific time lines and amounts of electricity that must be supplied by utilities through renewable energy sources, such as wind and solar. The RPS was conceived by the American Wind Energy Association (AWEA) and allies such as the Union of Concerned Scientists as a federal policy that would take the place of the Public Utility Regulatory Policies Act (PURPA) in helping to drive renewable energy markets, but when Congress did not act quickly in implementing an RPS on a national level, a growing number of states embraced the policy.

Most of the United States, particularly in the Upper Midwest and Great Plains, has an abundance of free wind just waiting to be tapped. Many developers have approached farmers and ranchers about securing access to their land to build turbines. Before this, most farmers and ranchers were known to curse the wind. With lucrative offers from wind farm developers to rent small parcels of land for individual turbines, farmers are now more inclined to praise it. "Basically, they're paying me to let the wind blow," Lake Benton, Minnesota farmer Conrad Schardin said to the *New York Times* in November 2000.[2] According to the *Times*, Lake Benton farmers at the time could earn as much as $2,000 a year per turbine on their property, and yet still use more than 95 percent of the land around the turbine's base for raising crops and grazing cattle.

Since there were no significant domestic turbine producers left in the U.S. market by the mid-1990s, wind farm developers relied largely on imports from Europe, especially Denmark's Vestas Wind Systems, Bonus Energy and NEG Micon (merged with Vestas in 2003).[3] However, in the late 1990s, Tehachapi, California-based Enron Wind surfaced in the wind energy business and began competing with traditional European manufacturers in the United States. Enron built its operations off acquisitions of established firms, such as Zond Energy Systems in the United States, and Tacke Windenergie in Salzbergen, Germany. The company amassed about 1,600 employees worldwide. In February 2002, General Electric bought Enron Wind after the Houston, Texas-based parent Enron Corp. went bankrupt. General Electric gave its new wind energy unit a boost by making available its vast pool of engineers.[4] Exam-

A shop floor view of Clipper Windpower's 330,000 square foot manufacturing facility in Cedar Rapids, Iowa. *Courtesy of Clipper Windpower, Carpinteria, California.*

The benefits of wind farm development spill over to the wider community. In 2004, the AWEA noted that the 108-turbine Colorado Green Wind Project in Prowers County, located in the southeastern part of the state, drew at its height about 400 workers from around the country to build access roads, pour foundations and set up electrical infrastructure.[19] Today, these turbines continue to supply local and county tax revenues, which help support schools, hospitals and public activities. A study by the Renewable Energy Policy Project in 2004 found that boosting U.S. wind energy installations to about eight times the 2004 level could create 150,000 manufacturing jobs across the United States. The study identified manufacturing activities for wind turbine components, such as ball and roller bearings, gearboxes, generators, transformers, power electronics, blades, and towers, in 25 states. These types of jobs are particularly welcome to states that have lost manufacturing to overseas competition. The Renewable Energy Policy Project also used the North American Industrial Classification System codes for 20 basic wind turbine components to find that 16,163 firms engaged in manufacturing activities related to these components.[20] Local institutes are springing up around the country to promote wind energy-related jobs. The West Texas Wind Energy Consortium, for example, encourages communities to focus on the retention and expansion of wind energy businesses, workers and other industries. "The tower wouldn't be standing there very long without all these parts," Greg Wortham, executive director of the consortium, told the *San Angelo Standard-Times*. "Some company somewhere is making these parts, and it is part of the wind energy."[21]

Knight & Carver technician Elizabeth Coronado puts finishing touches on a large turbine blade at the company's National City, California plant in January 2007. *Courtesy of John Freeman of Knight & Carver, National City, California*.

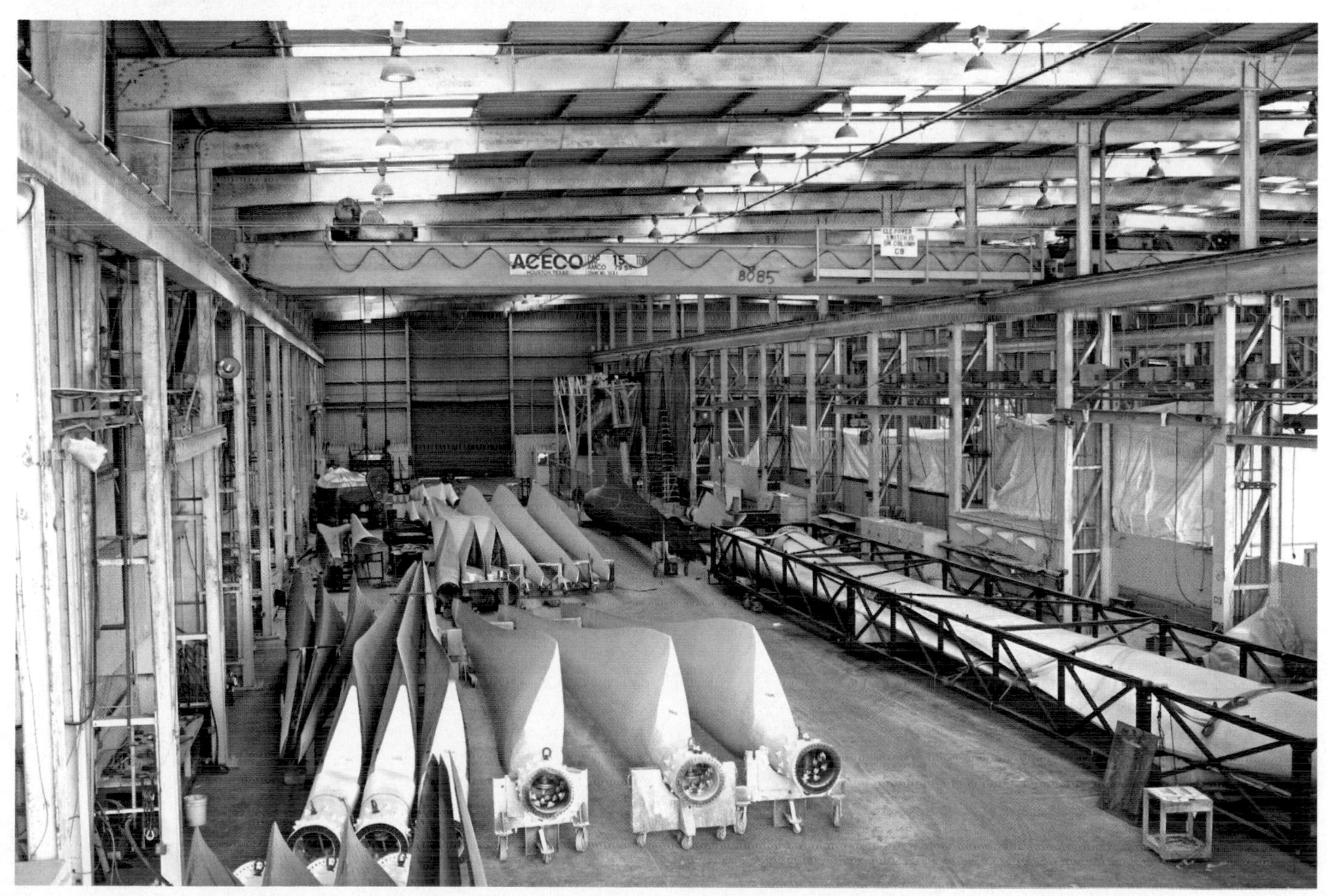


An overhead view of blade repair and production in progress at Knight & Carver's National City, California plant in October 2006. *Courtesy of John Freeman of Knight & Carver, National City, California*.

Many former industrial properties are becoming ideal locations for wind farms. In June 2007, eight Clipper Windpower 2.5 megawatt turbines were built on 30 acres of the former Bethlehem Steel plant in Lackawanna, New York, just south of Buffalo along the shore of Lake Erie. The $40 million Steel Winds Wind Farm was developed in partnership by UPC Wind and BQ Energy with guidance from the New York Department of Environmental Conservation Brownfield Cleanup Program, and it will provide power for about 6,000 western New York homes. "Where Bethlehem Steel once supported an earlier industrial revolution, today the Steel Winds project is bringing new jobs and clean energy technology to the Lake Erie region," said Paul Gaynor, UPC Wind president and chief executive officer, at the time of the wind farm's completion.[22] Besides being the first U.S. wind farm to use Clipper's 2.5 megawatt Liberty turbines, Steel Winds was considered the first "urban" wind farm in the United States and the first on the U.S. side of the Great Lakes.[23] BQ Energy has its eyes on the former Fresh Kills landfill on Staten Island, New York as another possible site well suited for wind turbines.[24] In the mountains of the Mid-Atlantic states, communities have become increasingly interested in converting former coal mining sites into wind farms. The ridge-tops of these sites are relatively flat from the mining. The Department of Energy's Office of Energy Efficiency and Renewable Energy estimates the Appalachians north of North Carolina to southern New York have wind speeds ranging from 12.5 to 15.7 mph, and the AWEA considers Pennsylvania to be the state with the region's most productive winds, followed by New York.[25]

The challenge for some wind rich areas of the United States is having sufficient infrastructure to deliver wind-produced electricity efficiently to the nation's power grid. North Dakota has been called the "Saudi Arabia of wind," because the state could theoretically produce enough wind-generated electricity for a fourth of the U.S. electricity demand. However, North Dakota's sparse population means that wind farm developers struggle to secure access to adequate transmission lines. After years of debate, the Federal Energy Regulatory Commission (FERC) on April 19, 2007 approved a new California Independent System Operator transmission policy. AWEA said the policy solves the "chicken and egg" problem long associated with wind power development where no wind farms are built unless there is transmission, and often no transmission is built unless there are wind farms already in place. The FERC ruling essentially supports that transmission should be financed and built as long as it is clear that there is a large energy resource to be tapped and that there is some financial commitment on the part of the generating companies to eventually develop wind projects in the area. Some states, such as Texas, Colorado, Minnesota and California, have implemented similar policies at the state level. With the FERC decision, this policy can now be pursued by every transmission provider in the country with a FERC jurisdictional transmission tariff. "FERC's unanimous decision is a breakthrough step toward getting transmission nationally into renewable resource rich areas and facilitates achievement of renewable portfolio standard goals," said Hal Romanowitz, president of Oak Creek Energy in California and an AWEA board member.[26] In early 2007, Minnesota Power Cooperative and Otter Trail Power Company announced the development of North Dakota's largest wind farm to date at Langdon. The wind farm will generate 159 megawatts at peak output through the use of 106 1.5 megawatt General Electric turbines. The two utilities will upgrade 35 miles of existing transmission line between Langdon and Hensel to deliver wind-generated electricity to the high-voltage transmission network. The wind farm will be operational by early 2008.[27]

Clipper's Liberty turbine is the largest manufactured in the Americas in 2007. On an annual basis, one unit can generate enough clean electricity to power about 800 average American homes. *Courtesy of Clipper Windpower, Carpinteria, California.*

More attention will be placed on developing ever-larger utility-scale wind farms throughout the United States. In 2007, the U.S. wind energy industry installed more than 5,244 megawatts of new power generation, an investment of more than $9 billion, for a total capacity of more than 16,818 megawatts, or enough to serve about 4.5 million average American homes. According to the AWEA, the top five states with the most wind energy installed by capacity (megawatts) in 2007 are:

1.) Texas – 4,356 megawatts
2.) California – 2,439 megawatts
3.) Minnesota – 1,299 megawatts
4.) Iowa – 1,273 megawatts
5.) Washington – 1,163 megawatts[28]

The largest wind farms operating in the United States by 2006 were:

1.) Horse Hollow, Texas (736 megawatts) – FPL Energy
2.) Maple Ridge, New York (322 megawatts) – PPM Energy/Horizon Wind Energy
3.) Stateline, Oregon and Washington (300 megawatts) – FPL Energy
4.) King Mountain, Texas (281 megawatts) – FPL Energy
5.) Sweetwater, Texas (264 megawatts) – Babcock & Brown/Catamount[29]

In 2007, FPL Energy remained the dominant developer and operator of wind farms in the United States with the goal to add between 8,000 and 10,000 megawatts of new wind projects to its portfolio by the end of 2012. In 2007, FPL Energy had more than 14,000 megawatts of wind projects in various stages of development, covering more than one million acres of land. Between 2000 and 2006, the company added more than 3,500 megawatts to its wind portfolio.[30] Other large firms to enter the business since 2005 were AES, BP, Goldman Sachs, and John Deere. Similarly, two-thirds of the more than 3,000 megawatts of new wind installations in 2007 were erected in Texas.[31] The expansion in Texas won't slow down much before 2012. In July 2007, Shell WindEnergy and Luminant, a subsidiary of TXU Corporation, announced their intention to build a 3,000-megawatt wind farm in the Texas Panhandle.[32] Just before this announcement, T. Boone Pickens, a billionaire oil tycoon, announced his intention in June 2007 to build the world's largest wind farm in West Texas. His company, Mesa Power, plans to build as many as 2,000 wind turbines, some big enough to generate 2.5 megawatts, across as much as 200,000 acres spanning four counties. The $6 billion project, which Pickens hopes to start building in 2010, would produce up to 4,000 megawatts of electricity, or more than five times the power of the largest wind farm at Horse Hollow, near Abilene, Texas, in 2007.[33]

President George W. Bush, who initiated one of the most effective state-based renewable energy portfolio programs when he was governor of Texas, challenged the country as a whole in February 2006 when he released his Advanced Energy Initiative. The initiative described his vision for increasing the energy efficiency of businesses and homes through 2030 by greatly expanding the use of renewable energy sources to reduce the dependency and supply constraints on natural gas and other fossil fuels. The president's initiative also noted that wind-rich areas of the country could conceivably meet 20 percent, or more than 320 gigawatts, of U.S. electricity demand in the next 25 years. The AWEA believes there is more than enough wind to meet the president's 2030 renewable energy target, but a number of factors will need to be met to make the vision a reality. These include the maintenance of stable long-term government policies that promote wind energy development, such as multiyear production tax credits and a national renewable standards portfolio; elimination of supply chain constraints for turbine manufacturing and wind farm construction; expansion and enhancement of transmission lines to deliver power to the national grid; streamlined wind farm permitting regimes; a more flexible electric system to effectively absorb the variability of wind as an energy source; and ongoing efforts by the industry to keep wind power economically competitive with fossil fuels and other renewable energy sources.[34] In October 2007, the AWEA in cooperation with the Department of Energy and National Renewable Energy Laboratory released its technical report analyzing the "20% Wind Vision." AWEA also launched a five-year Action Plan that will lay the foundation for the 20% Wind Vision report by systematically attacking each of the key constraints facing the wind industry.

Views of the three John Deere Co.-financed 2005 wind projects which straddle the state line between the Texas and Oklahoma panhandles. The projects each consist of eight Suzlon S64 1.25 megawatt turbines, combined for a total of 30 megawatts of output. *Courtesy of Suzlon Wind Energy Corporation, Chicago, Illinois.*

Canada marked a record in 2006 for installed wind energy capacity. The country added 775 megawatts worth of wind power, boosting its overall capacity to 1,460 megawatts. According to the Global Wind Energy Council, another 21 wind energy projects were commissioned in 2006, which should add another 2,500 megawatts of wind power in the next few years.[35] The Canadian Wind Energy Association forecasts a minimum of 10,000 megawatts generated from wind farms scattered across the country's six provinces by 2015.[36] Like North Dakota, Quebec is considered the "Saudi Arabia of wind" in Canada. In 2007, Quebec was home to 321 megawatts of wind power, and by 2015 will reach an estimated 4,000 megawatts of the country's total forecasted wind energy output.[37] Canada's wind energy business is not immune to the challenges experienced by the industry in other parts of the world and is largely dependent on tax breaks and other favorable government incentives for its viability. Canada offers a 1 cent (Canadian) per kilowatt hour production incentive payment for the first 10 years for qualified wind farms. This has attracted the attention of wind farm developers and turbine manufacturers from the United States and Europe to consider operations in Canada.[38] Adding to these incentives, businesses, municipalities, institutions and organizations are eligible to apply for funding under the Canadian government's ecoENERGY for Renewable Power program, announced in January 2007. The initiative provides C$1.48 billion to increase Canada's supply of electricity from renewable sources such as wind, biomass, low-impact hydro, geothermal, solar photovoltaic, and wave energy. The government expects the program to encourage production of 14.3 terawatt hours of new electricity from renewable energy sources, enough electricity to power about 1 million homes. Canada's Minister of Natural Resources, Gary Lunn, noted: "Canada has huge reserves of fossil fuels, but we have to manage those resources wisely to protect the health of our economy and of our environment for the long term."[39] Kettles Hill Wind Energy, a subsidiary of Creststreet Kettles Hill Windpower LP, was the first company to receive funding under the ecoENERGY program in the amount of C$16.5 million. In 2007, Kettles had completed construction of a 63-megawatt wind farm located near Pincher Creek, Alberta. The project will produce up to 200,000 megawatt hours of power each year, enough to power more than 27,000 average homes. Kettles also operates the 54-megawatt Mount Copper wind farm located near Murdochville, Quebec, and the 30-megawatt Pubnico Point wind farm located near Yarmouth, Nova Scotia. Kettles plans to build a 180-megawatt Dokie wind farm located near Chetwynd, British Columbia in 2008.[40]

A dozen of the 47 General Electric 1.5 megawatt turbines at the Soderglen wind farm in Southern Alberta, Canada. *Courtesy of McKibben & Bailey, Calgary, Canada/Global Wind Energy Council, Brussels, Belgium.*

Mexico is also starting to put its wind resources to work. It is estimated that 30,000 megawatts of wind power is available in the Oaxaca region of the country. The LaVenta II wind farm in Oaxaca was completed in late 2006 by Spanish consortium Iberdrola-Gamesa and generates 83.3 megawatts. (LaVenta I was only a test facility.) LaVenta III will add another 101 megawatts of power to the wind farm and should be completed by 2009. Many companies will be involved in the third phase, such as Gamesa, Iberdrola, EDF-EN, Union Fenosa, GE Wind, Clipper Windpower, and Endesa. The Mexican Wind Energy Association reported in 2006 an installed wind capacity of 87 megawatts, and they forecasted that between 2006 and 2014 at least 3,000 megawatts of wind power will be generated. To add impetus to the effort, the Mexican congress approved a renewable energy use law in December 2005, which requires that 8 percent of the country's power be met by renewable energy sources, excluding large hydro projects. A major problem for Mexico's wind farm developers has been suitable access to transmission lines. The World Bank granted Mexico a $70 million loan to boost its grid-connections for renewable energies.[41]

Gamesa G52-850 kilowatt turbines at the Comisión Federal de Electricidad's La Venta II wind farm in the Mexican state of Oaxaca. *Courtesy of Tomás Soto and Miguel Lagos of Iberdrola Ingenieria y Construccion, Madrid, Spain/Global Wind Energy Council, Brussels, Belgium.*

# Chapter Ten
# Global Spread

Not to be underestimated, Asia seeks to exploit its wind energy potential in the years ahead. Wind farms have sprung up across the continent since the early 1980s. Many of Asia's most populous countries, namely India, China, South Korea and Japan, import large amounts of coal and oil to keep their power plants operating and their industrial engines humming. The increased costs of energy imports and pollution resulting from burning fossil fuels have encouraged these countries to introduce renewable energy portfolios, which in large part favor wind. Other emerging Asian markets, such as Vietnam, Sri Lanka, and the Philippines, are expected to increase their wind energy investments in the coming years. According to the Global Wind Energy Council, the total installed wind energy capacity on the Asian continent should reach 29 gigawatts by 2010, up from 10.7 gigawatts in 2006.[1] The development of wind farms has also helped to stimulate a regional market for turbine manufacturing which should remain on an upswing for the next decade.

By 2005, India surfaced as the dominant player in Asia, both in wind power usage and manufacturing. The Indian government began to take interest in renewable energy in the early 1980s to encourage energy diversification. The country is estimated to have a wind capacity of 65,000 megawatts.[2] India has since implemented incentives to attract and encourage the development of wind farms. For example, India's Electricity Act of 2003 created state electricity regulatory commissions to promote renewable energy such as wind. The legislation also introduced Renewable Portfolio Standards, which require utilities to buy a percentage of their electrical power from renewable energy sources. India implemented a production tax credit program for nonresident Indian firms to attract foreign investment in wind farms.[3] More than 97 percent of the wind power units operating in India today are controlled by private sector interests.[4] Half of India's wind farms are located at Tamil Nadu in the southern part of the country. Other states which are building wind farms include Maharashtra, Gujarat, Rajasthan, Karnataka, West Bengal, Madhya Pradesh and Andhra Pradesh. The Indian government plans to generate 10,000 megawatts from wind power by 2010.[5]

Suzlon is India's top wind turbine manufacturer. *Courtesy of Suzlon Wind Energy Corporation, Chicago, Illinois.*

Indian billionaire Tulsi Tanti has become the leading personality of India's turbine manufacturing sector. While operating a textile plant in the early 1990s, Tanti became frustrated by his excessive energy bills and got the idea to erect a wind turbine to reduce the plant's energy costs. His interest in wind energy took off and he eventually sold all his other businesses to concentrate solely on wind turbine manufacturing.[6] Tanti's company, Suzlon Energy Limited, is based in Pune, near Mumbai. The company's main manufacturing plants in India are located in Daman, Pondicherry, Maharashtra and Gujarat. Suzlon's turbines range in size from 0.35 megawatts up to 2.1 megawatts. The company also builds its own rotor blades, generators, gearboxes, control systems, and tubular towers. Suzlon has established numerous projects and overseas production facilities, including locations in United States, Europe, South Korea and China. In 2006, Suzlon was ranked fifth largest wind turbine supplier in the world.[7] Other large turbine manufacturers operating in India are Enercon, Vestas Wind Technology India Pvt. Ltd., and Pioneer Asia Wind Energy Group.[8]

A field engineer inspects a turbine at the Guangdong Nan'ao wind farm in the southern Chinese province of Guangdong. *Courtesy of Greenpeace China, Hong Kong/Global Wind Energy Council, Brussels, Belgium.*

China became the second largest wind energy producer in Asia by 2006. The Global Wind Energy Council noted that China reached 2,604 megawatts of wind power that year. China's goal is to construct enough wind turbines to generate 30 gigawatts by 2020, which would cover 1.5 percent of China's electricity demand.[9] China erected its first wind farm at Rongcheng in the northeast province of Shandong in 1986.[10] By the end of 2005, China had a total of 62 wind farms, the majority of which were located in Liaonin, Guangdong, Xinjiang, and Inner Mongolia.[11] (Hong Kong Electric Company erected its first wind turbine on Lamma Island in early 2006. The turbine is capable of generating 800 kilowatts of power.) China's Meteorology Research Institute estimates that the country's overall exploitable land-based wind power is about 253 gigawatts.[12] The Chinese government has developed a number of programs to stimulate and maintain its burgeoning wind power sector. These programs include tax incentives for developers, fixed standardized electricity rates, and domestic manufacturing requirements.[13] The government wants 70 percent of wind turbine components to be produced domestically. This has encouraged a number of major manufacturers, such as Vestas, Gamesa, Suzlon and General Electric, to set up factories in the country. More than 20 Chinese firms are involved in wind turbine production. These companies produced about 400 megawatts worth in wind turbine components in 2006.[14] The Chinese government has also developed regulations to support clean energy technology development and to promote power grid access and adequate payments for wind power producers. On January 1, 2006, a new Chinese law to promote the development of renewable energy technologies took effect. However, to compete with cheaper electricity generated from coal fired plants, China aims to develop mostly large-scale wind farms in high wind areas such as Inner Mongolia, Jiangsu, Hebei and Jilin.[15]

By early 2007, Taiwan installed more than 100 wind turbines on its west coast. These turbines receive strong northwesterly winds for six months of the year. The turbines are spread across 13 wind farms and generate about 420 million kilowatt hours a year, or enough power for about 105,000 households. The turbines were constructed by the Taiwan Power Company, Tien Lung Paper Company, and Infra Vest Wind Power Group, a German company which was the first to enter the Taiwanese market in 2000.[16] Taiwan Power began installing turbines in 2002 and will erect 200 units by 2010. The company plans to build another 564 units between 2010 and 2020 in the shallow waters off Taiwan's west coast and Penghu island group for a total capacity of 1,980 megawatts.[17] Taiwan's Bureau of Energy has called for 10 percent of the country's energy needs to be met through renewable sources by 2010, of which 80 percent will be generated by wind power.[18]

A wind turbine in the eastern part of Nanao Island, Guangdong Province, China. The stone in the foreground reads "Buddhism." *Courtesy of Greenpeace China, Hong Kong/Global Wind Energy Council, Brussels, Belgium.*

By August 2007, the Huitengxile wind farm in Inner Mongolia operated 100 turbines capable of generating 130 million kilowatts a year, making it the largest wind farm in China. The project will consist of up to 300 turbines once fully built. *Courtesy of the China Wind Energy Association, Beijing, China.*

Japan ranked eighth largest generator of wind power in 2006, but that statistic was far behind India and China in terms of actual output. Japan's leading wind areas are Tohoku and Hokkaido in the north and Kyushu in the south. The country's rough terrain poses limits to construction and frequent typhoons have toppled turbines. There is also hesitancy by Japan's electric companies to embrace wind energy because of concerns over power fluctuations on the grid.[19] However, the country has not given up on wind power. In April 2003, the Japanese government introduced a renewable portfolio standard with the goal to provide 3,000 megawatts in wind energy to the country's electrical supply by 2010.[20] Private-citizen groups in Japan have also funded and built their own utility-scale wind turbines. The Hakkaido Green Fund set up its first wind turbine in Hamatonbestu, Hokkaido in September 2001 and erected 10 more turbines by the end of 2006. The turbines stand at 197 feet (60 meters) high, each producing enough electricity for 1,200 households. Each turbine costs about 300 million yen (or about $2.61 million).[21] In addition, Japan has one of the world's largest manufacturers of wind turbines. Mitsubishi Heavy Industries announced in April 2007 that it plans to increase production of wind turbine generators from the current 400 megawatts per year to 1,200 megawatts per year by mid-2008 to respond to international demand, especially in the United States. The investment is worth 4 billion yen, the company said.[22]

South Korea's interest in wind energy became prominent after the 1997-98 financial crisis.[23] By 2006, the country installed 173 megawatts of wind capacity. The government plans to build 1,137 megawatts by 2010 and 2,250 megawatts by 2012. But like Japan, South Korea suffers from a lack of adequate roads and power grids that can reach into the windy mountainous areas of the country.[24] Despite this setback, the country has a thriving domestic turbine manufacturing base, which includes Hyosung, Unison, Aurorawind, and Junma Engineering. These companies build turbines in the capacity range of 750 kilowatts but have plans to expand development

Seven Gamesa G80 2 megawatt turbines at the Summit Wind Power farm in the industrial section of Kashima, Japan. *Courtesy of Sumitomo Corporation, Tokyo, Japan/Global Wind Energy Council, Brussels, Belgium*.

of units with output rates of up to 3 megawatts.[25] In 2007, Unison built South Korea's largest wind farm on the ridges of Taegwanryong in the Kangwon province. The site contains 49 turbines each capable of producing 2 megawatts, or enough together to power 50,000 homes.[26]

North Korea is also no stranger to wind energy, although on a much smaller scale than its neighbor to the south. The country received outside help in 1986 from Denmark to install some wind turbines on its west coast.[27] In 1998, five American engineers promoting renewable energies through the California-based Nautilus Institute received a $400,000 grant from the W. Alton Jones Foundation in Virginia to construct an 11.5 kilowatt wind system in the North Korean village of Unhari, about 30 miles north of Nampo on the country's west coast.[28] Seven small turbines were shipped from the United States and erected over a five-week period. The project's aim was to provide electricity to a medical clinic, kindergarten classroom, and village households. On Oct. 5, 1998, the system was switched on.[29] The U.S.-DPRK Wind Power Village Humanitarian Energy Project also involved about 50 North Korean engineers, technicians and laborers, meeting the Nautilus Institute's goal to encourage cooperation between the two countries.[30] By October 2000, the Nautilus Institute team made its third trip to Unhari to inspect the existing turbines, make necessary repairs, provide additional training, and install a mechanical windmill to provide drinking water.[31] While the project proved successful, overall political tensions between the United States and North Korea escalated to the point where the wind energy project had to be terminated in 2006. "We made a joint decision with our DPRK counterpart to dismantle the system because of this high-level context, but also because we could not raise the funds in the United States to sustain the appropriate level of maintenance to ensure safe operations of the plant in the village," said Peter Hayes, who led the project.[32] The North Koreans dismantled the turbines in 2006 and transferred them to storage.[33] In late 2006, the North Korean government expressed interest at an Asian Energy Security Workshop in building a 10-megawatt prototype wind farm by 2010, with the goal to build three large 100-megawatt wind farms by 2015.[34]

In 2005, Australia's progress with wind energy nearly came to a halt when the government decided not to extend its wind energy industry support program beyond 2006. Australia was a noted pioneer for its mandatory renewable energy target, which required the country's power utilities to increase their amount of energy intake from renewable sources, such as wind. Without this program, Australia's wind farm operators can't generate a profit, and it is difficult enough already to compete with the country's abundance of cheap coal.[35] The Global Wind Energy Council estimated that Australia's installed wind energy capacity for 2006 was at 817 megawatts.[36] But Australia's wind energy sector received hope in late 2006 when three states stepped up with their own mandatory renewable energy target programs. Victoria plans to receive 10 percent of all its electricity from renewable sources by 2016. South Australia, which has half of the country's installed wind power, will generate 20 percent of its electricity from renewable sources by 2014, and North South Wales has set a target of 15 percent by 2015.[37]

Wind turbines are popping up in the Middle East and Africa. In 2006, the Global Wind Energy Council estimated a combined installation of wind turbines for the regions of 441 megawatts.[38] Although Iran sits on vast oil and gas reserves, the country has become an ardent supporter of wind power in the Middle East. In 1999, Iran's Ministry of Energy formed the Renewable Energy Organization and the country erected its first wind farm – rated at 12 megawatts – in 2000. While these turbines were imported, the government is actively promoting domestic production of these units. In 2007, SabaNiroo Company remained Iran's sole turbine maker and the only producer in the Middle East. Wind energy production in Iran by 2006 reached 48 megawatts.[39] By March 2007, SabaNiroo installed and commissioned ten 300 kilowatt, ten 55 kilowatt, and twenty-two 660 kilowatt turbines in Manjil; twenty 660 kilowatt turbines in Binalood; and four 660 kilowatt turbines in Armenia (Pushkin Pass site).[40] North Africa's dominant wind power developers are located in Egypt and Morocco. Egypt's New and Renewable Energy Authority, formed in 1986, forecasts that about 850 megawatts in wind energy systems will be installed by late 2010. The country hopes to generate 14 percent of its national power through renewable energy by 2020, of which 7 percent, or 2,750 megawatts, will come from wind.[41] Egypt's three wind farm areas – Hurghada, Zafarana, and Gabal El-Zayt – are located in the Suez Gulf region.[42] In 2006, Morocco doubled its wind capacity to 124 megawatts. A new 60 megawatt wind farm, called Amogdoul, was built at Cap Sim. The site consists of 71 turbines. The country has approved construction of a wind farm at Tarfaya along its coast. The wind farm will start with 200 megawatts of output and increase to 300 megawatts, equipped with 850 kilowatt turbines. Construction is scheduled to begin at Tarfaya in 2010. Another 100 megawatt capacity wind farm site is set for Touahar, along with 13 other locations in the planning stages.[43] Within the African continent, there is plenty of discussion about the potential for wind energy. In early 2007, the Namibian government issued a license to Aeolus Power Generation Namibia, a joint venture between Dutch firm Aeolus Association and Namibian company United Africa Group, to install turbines in the country's coastal region.[44] Another African wind project announced in 2007 is set for Marsabit, Kenya. The wind farm will be developed by British firm African Clean Energy and Wind Flow, a Kenyan company. Like many African countries, Kenya yearns to develop quicker, cleaner ways to deliver power to its people. Only 8 percent of Kenya's population in 2007 had access to electric power. Some individual firms in the country have embraced wind power on a small scale. Mobile service provider Safaricom Ltd., for example, uses 40 kilowatt wind turbines to power remote base stations throughout the country.[45]

Row of Saba Niroo Co. turbines at Iran's Binalood wind farm. *Courtesy Global Wind Energy Council, Brussels, Belgium.*

Vestas V42 600 kilowatt turbines at the Koudia Al Baida-Abdelkhalek Torrés wind farm near the city of Tetouan on the northern coast of Morocco. The wind farm sits less than 10 miles from Spain across the Straight of Gibraltar. *Courtesy of Centre de Développement des Energies Renouvelables, Marrakech, Morocco/Global Wind Energy Council, Brussels, Belgium.*

A chaotic view of numerous types of wind turbines in the San Gorgonio Pass near Palm Springs, California. *Courtesy of Keith Higginbotham, Long Beach, California*.

For some people there is a nagging feeling of chaos when encountering a wind farm, especially when the winds are turning a majority of the turbine blades at once. A 465-foot-tall turbine – from tower base to the tip of the tallest blade – dwarfs the average 125-foot-tall electrical transmission tower. Turbines may also cause a landscape phenomenon called "shadow flicker," a strobe-light-like effect which occurs when the light from the setting or rising sun passes through the rotating blades. Some individuals who live in close proximity to the turbines claim that shadow flicker induces headaches and depression.[1]

Suzlon S64 1.25 megawatt turbines, part of the John Deere Co.-financed wind projects in 2005, blend into the environment along the Texas and Oklahoma panhandles. *Courtesy of Suzlon Wind Energy Corporation, Chicago, Illinois*.

Wind energy trade associations say these claims are "myths," perpetuations of the anti-wind lobby, or individuals who they refer to as NIMBYs ("Not-In-My-Backyard"). In 2005, the American Wind Energy Association (AWEA) helped form a broad coalition of organizations from the environmental, agricultural, business, health, social justice, faith and academic communities to support wind energy. The goal of the coalition, known as Wind Energy Works!, is to "counter myths with facts."[2] Myths or not, lawmakers worldwide generally listen to the anti-wind lobby as much as they do to wind energy proponents. A split in opinions about the benefits of wind energy among politicians, federal and state agencies, and community activists may bog down wind projects for years. Such was the case of a proposed wind farm in Garrett County, Maryland, which received state approval in mid-2007. Some studies have found that opposition to wind turbines persists simply due to the public's lack of knowledge about wind energy. A research group led by St. Andrews University in the United Kingdom discovered a phenomenon which they termed "reverse-nimbyism." This occurs when individuals once opposed to wind energy live among the turbines and come to realize their environmental benefits.[3] Paying more for fossil fuels has turned many people's attention to wind. A 2007 Berlingske Tidende/Gallup poll found that 44 percent of Danish citizens were willing to support construction of more turbines, even if it meant a 10 percent increase in the price of electricity.[4] The National Academics in the United States suggested improved public opinion of wind could be attained by not overcrowding the landscape with small turbines.[5]

Still some wind projects struggle to fit in, especially those whose turbines have been linked to bird and bat kills. One of the oldest and most troubled wind farm regions in this regard is California's Altamont Pass. The pass consists of about 7,000 turbines spread across 80 square miles of eastern Alameda and Contra Costa counties. Altamont Pass is considered one the largest, most highly concentrated wind farm areas in the world. This California region is also home to one of largest raptor populations. The unsettling relationship between these raptors and Altamont's whirling turbine blades has been known since the early 1980s. U.S. Windpower, once a prominent Altamont wind farm developer, tried to diminish bird collisions with its turbines in the 1980s by testing noise makers and painting the blades in a "disruptive pattern of red and blue" in the hope of turning the birds away. The company also set up a fund for rehabilitators to nurse injured birds.[6] In the late 1980s, the Department of Energy's National Renewable Energy Laboratory studied bird behavior at Altamont as it related to turbine blade and tower design, turbine clusters and locations, and landscape management practices to thwart bird kills. Two decades later, however, Altamont is still considered the most deadly site for raptors. According to a 2004 report by the California Energy Commission, between 881 and 1,300 raptors are killed annually at Altamont. These raptors include golden eagles, red-tailed hawks, American kestrels and burrowing owls. Many types of raptors are protected under several national laws, including the Migratory Bird Treaty Act, Bald and Golden Eagle Protection Act, Endangered Species Act, and the Marine Mammal Protection Act (for offshore wind farms). Song birds within the Altamont area are also occasionally hit and killed by wind turbine blades, raising the annual bird death toll to between 1,766 and 4,721. The 540-page report is based on five years of research involving 2,548 wind turbines in Altamont.[7]

Cattle graze comfortably among the whirling blades of these Suzlon S64 1.25 megawatt turbines along the Texas-Oklahoma state line. *Courtesy of Suzlon Wind Energy Corporation, Chicago, Illinois*.

The California Energy Commission report made numerous recommendations to help mitigate bird kills at Altamont, such as replacing small turbines with fewer, much larger models, a procedure in the business known as "repowering." "The effort that the repowering program will have on bird mortality is unknown; however, the research presented in this report suggests that repowering may reduce mortality, especially if turbines are installed on the tallest practicable towers," the report said.[8] Other mitigating efforts suggested are to end rodent control programs; relocate selected, highly detrimental turbines; move rock piles away from turbines; develop anti-burrowing measures around turbine pads; remove non-operating turbines; install bird migration monitoring equipment; and retrofit dangerous power poles. The report also recommended reducing sloped cuts in access roads, preventing cattle from grazing under turbines, installing flight diverters, and painting blades. In addition, wind farm operators may install perch guards on their turbines, provide alternative perches, and barricade the turbine rotor planes. Commission researchers believe that if all these measures are implemented, bird kills in Altamont could be reduced by 40 percent.[9]

Raptors and song birds aren't the only victims of turbine blade collisions. Two recently built wind projects on the East Coast have been associated with killing bats. The wind farms are located at Backbone Mountain, West Virginia, and Meyersdale, Pennsylvania. Bat deaths were noticed at the 44-turbine Mountaineer Wind Energy Center on Backbone Mountain shortly after starting operations in 2003. Similar observations were recorded at the 20-turbine Meyersdale site, which also started in 2003. Researchers aren't completely sure why bats are colliding with these particular turbines, but some believe that they may be attracted to the presence of insects within the cleared ridgelines where the turbines are installed.[10]

Some environmental groups and national lawmakers have accused the U.S. Fish and Wildlife Service of not pursuing criminal action against wind turbine operators who knowingly violate wildlife protection rules.[11] Instead of issuing penalties, the agency has generally preferred to work with turbine operators to find ways to reduce bird and bat kills.[12] Fish and Wildlife has also stepped up efforts to work with local and state officials on how to identify wildlife concerns when assessing wind energy projects. In 2002, the agency formed the Wind Turbine Siting Working Group to develop comprehensive national guidelines for locating and building wind farms. Guidelines were proposed in 2003, but the method used by the agency to form the committee was questioned for possible violations of the Federal Advisory Committee Act, and the committee was subsequently disbanded. On March 13, 2007, Fish and Wildlife proposed the Wind Turbine Guidelines Advisory Committee, which is composed of members from selected wind energy and conservation associations, consulting firms, Indian tribes, and local, state, and federal agencies. The committee will provide advice and recommendations to Fish and Wildlife on developing measures to reduce the impact of landside wind farms on wildlife and their habitats.[13] Fish and Wildlife plans to use the results from this committee's work to develop a "national template" for voluntary bird protections in the wind energy industry.[14] Yet these protections may not be enough to stop some wind farm developers and local and state regulators, who are under pressure to meet mandated renewable energy portfolio and tax credit deadlines, from disregarding protections for birds and bats. Some environmental groups want Congress to develop stricter wildlife protections, both for pre- and post-construction, as a condition for wind farm developers to receive the federal Production Tax Credit and other subsidies.[15] "The bottom line is that we cannot allow ourselves to wholeheartedly embrace wind energy at every location where a strong wind blows, without first evaluating this technology in its entirety and having in place a responsible regulatory framework," said Rep. Nick J. Rahall II, chairman of the U.S. House Committee on Natural Resources, during May 1, 2007 hearing.[16] In favor of balance, the National Academies in a report found no evidence that existing wind farms are causing significant changes in U.S. bird populations. They warned, however, that studies should be conducted to evaluate the possible wildlife impacts for proposed wind projects and that follow up studies after construction should also be performed.[17]

Wind farm developers abroad are under similar pressure to implement bird protections. In early 2007, a proposed 12-turbine wind farm in Turkey was put on hold by the government over concerns that it was too close to a known migratory bird flight path.[18] In Spain, real-time radar has been implemented to monitor raptor migrations through wind farms. The World Bank, as a condition of issuing its development loan, requires similar technologies to be installed at wind projects in Mexico's Oaxaca region.[19]

Some high-tech experiments with raptor tracking technologies are taking place in the United States. In 2006, National Aviary researchers in Pittsburgh released two golden eagles equipped with tiny radio transmitters along the Alleghany Front ridge in Central Pennsylvania. The purpose of the experiment is to track the raptors' migration patterns and provide better information to help prevent wind farm developers and state regulators from allowing turbines in bird migration paths. As many as 10 golden eagles are expected to participate in the program in 2007.[20]

Wind power supporters argue that the bird and bat mortality issue is being overplayed and misguided by anti-wind groups. There are plenty of examples in recent history where industrial processes and accidents have wiped out tens of thousands more birds than all the wind farms in the country put together. A 1989 study reported that more than 500,000 birds are killed annually in oilfield waste pits.[21] The tanker *Exxon Valdez*, which spilled about 11 million gallons of crude oil into the Prince William Sound after striking a reef on March 24, 1989, killed an estimated 250,000 sea birds and 250 bald eagles. Hundreds of millions of birds are killed annually from collisions with vehicles, windows, communications towers, and aircraft engines. Countless birds also experience deadly encounters with billowing smoke stacks, pesticides and house cats. Mike Tidwell, director of the Chesapeake Climate Action Network, noted in a letter to the editor in the Feb. 6, 2005 *Washington Post* that about 490,000 acres of Appalachian Mountain forests throughout West Virginia, Virginia, Kentucky and Tennessee have been turned into a "moonscape" from coal mining in the past decade. "How many Appalachian bats were exterminated in the past dozen years as a byproduct of this process? Surely millions," Tidwell wrote.[22] In recent years, the AWEA has worked with numerous wildlife conservation groups to gain a better understanding of bird and bat habitats. These initiatives include the National Wind Coordinating Committee, Bats and Wind Energy Cooperative, and Kansas State University's Grassland Shrub – Steppe Species Collaborative. AWEA Executive Director Randall Swisher, in a 2005 statement about bats and wind turbine blades, declared: "This industry believes that bats and wind turbines can and must coexist, and is working with stakeholder groups and experts to understand the issue and try to find ways to avoid or at least reduce collisions... We have nothing to hide."[23]

Chapter Twelve

# Growing Up

By 2005, many wind turbine manufacturers had set their sights on building higher output units. Two and 3 megawatt turbines are widely available for installation. By late 2006, production of 5 megawatt turbines for offshore wind farms emerged. These modern giants of land and sea supersede the first and second generation turbines from the 1980s and early 1990s. A single 1.5 megawatt turbine with a rotor diameter of 230 feet (70 meters) alone replaces twenty-three 65-kilowatt machines with rotor diameters of 51 feet (15.5 meters).[1] While many early glitches associated with turbine blade control systems, gearboxes, and towers have been eliminated, the next generation of large turbines poses a new set of technical challenges to wind project developers.

600'
500'
400'
300'
200'
100'
0'

**1996**

The 550-kW Zond Z-40, commonly installed in 1996, has a 131-ft (40-m) hub height and is as tall as a 12-story building.

**2006**

The GE 1.5-MW wind turbine, commonly installed in 2006, has a 275-ft (84-m) hub height and is almost as tall as the Statue of Liberty, which is 305 feet (93-m) tall from the ground to tip of torch.

Wind turbine sizes continue to increase. *Courtesy of the National Renewable Energy Laboratory, Golden, Colorado.*

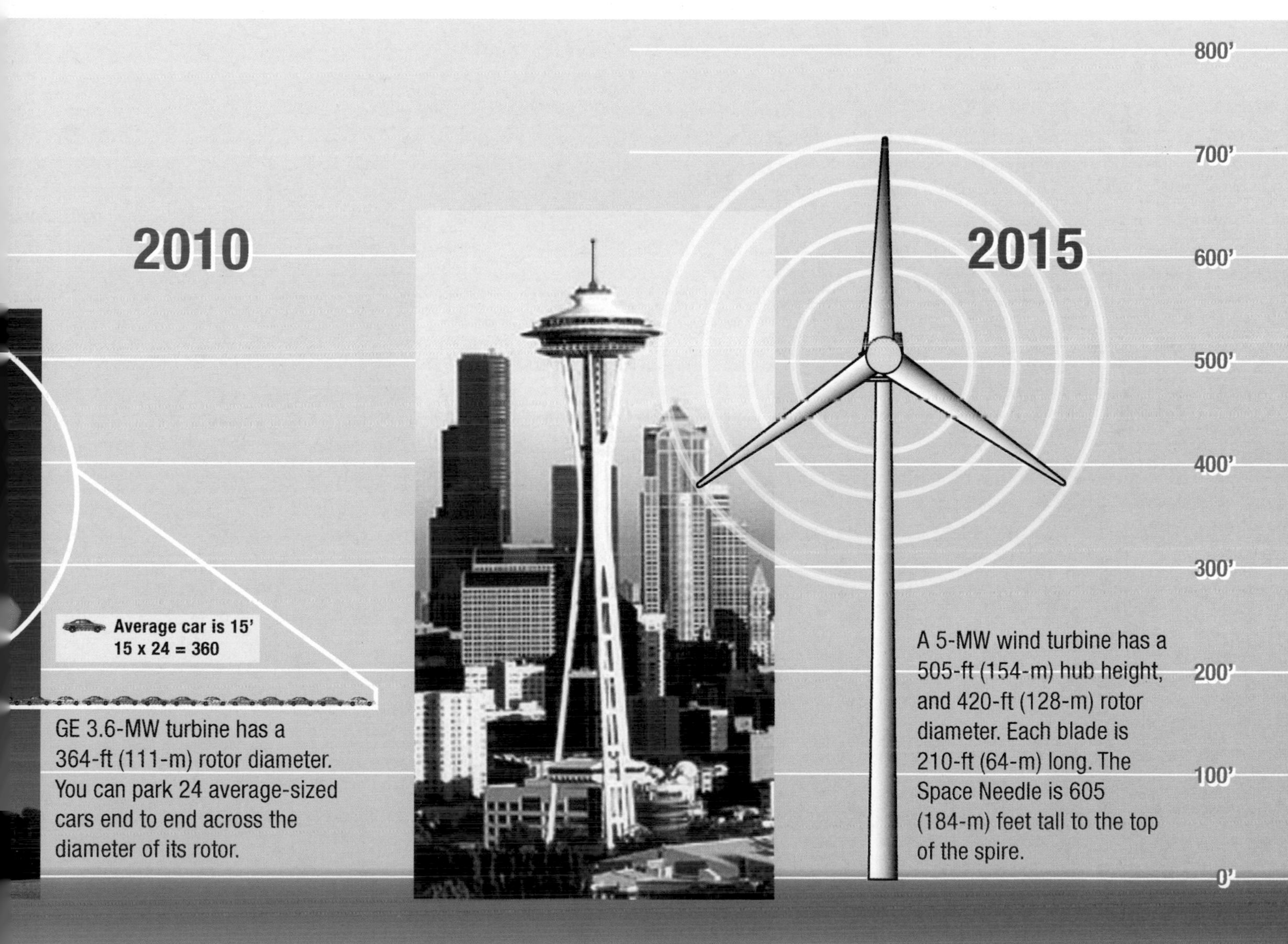

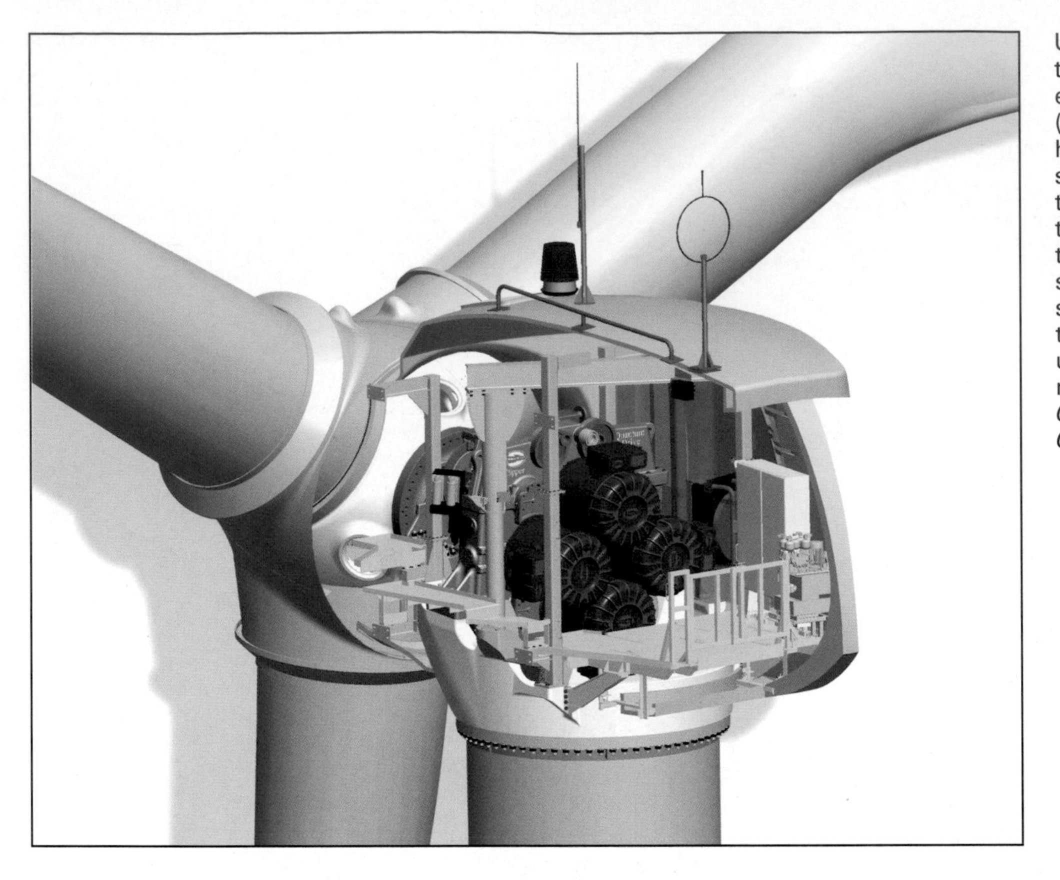

Unique in the world of wind turbine design, Clipper Windpower's "Quantum Drive" powertrain (c. 2007), a compact, two stage helical distributed load path design, uses four high-speed shafts that distribute torque loads from the rotor mainshaft to the generators, enabling torque loads to be split by a factor of eight at low speed stage factor and a factor of two at high speed stage, distributing loads, minimizing maintenance, and extending turbine life. *Courtesy of Clipper Windpower, Carpinteria, California.*

Blade research and development has taken center stage in the design of large turbines. Blades are considered the most important turbine components. They capture the wind's energy and are the source of most of the system loads.[2] Blade testing is also required to meet wind turbine design standards, reduce machine costs, and minimize the technical and financial risk of deploying mass-produced wind turbines. The challenge for engineers is how to build larger, longer blades without increasing the weight per unit of energy produced. Blades are traditionally fashioned from fiberglass using processes developed by the boat-building business. Some lightweight materials, such as carbon, are now used in the construction of the largest turbine blades. Most innovations, however, are taking place in the blade manufacturing process itself, such as resin infusion technologies.[3]

The early production stages of the STAR low-wind-speed blade in Knight & Carver's production plant in National City, California in December 2006. *Courtesy of John Freeman of Knight & Carver, National City, California.*

The U.S. government has invested millions of dollars in recent years to further research into larger turbine blades. This research has been carried out by the Department of Energy's Sandia National Laboratories and National Renewable Energy Laboratory with the goal to "look for ways to achieve mass production savings, use less expensive and/or higher quality materials, and produce more efficient structural and aerodynamic designs."[4] By 2006, the National Renewable Energy Laboratory's National Wind Technology Center had tested blades up to 148 feet in length for turbines ranging in size from 1.5 megawatts to 2.5 megawatts.[5] Until recently, the National Wind Technology Center near Boulder, Colorado was the only test facility in North America capable of full-scale testing of megawatt-size turbine blades. However, the move by some manufacturers toward production of 5 megawatt offshore turbines outstripped the laboratory's ability to test blades nearing 200 feet in length. In an effort to expand the federal government's blade research capabilities, the Department of Energy reached out to the scientific community for help to design, build and operate new facilities to test the next generation of wind turbine blades. Bids were sought and two sites were selected in June 2007. The Commonwealth of Massachusetts Partnership will construct a test facility at Boston Autoport in Boston Harbor, while the Lone Star Wind Alliance will set up a similar operation at Ingleside, Texas. Both organizations are consortiums of state and academic institutions.[6] Transportation was a key factor in the government's selection of the two sites. The Boston Autoport's location provides immediate access to substantial offshore wind resources, truck access, a rail spur and a 1,200-foot dock for offloading blades from ocean vessels. The Ingleside site benefits from its proximity to active shipping routes along the Gulf Coast and provides sufficient access to developing wind energy markets in Texas and the Midwest. The facilities are expected to be operational in 2009.[7] The ultimate goal for the two new facilities is to test blades up to 330 feet long. "These two testing facilities represent an important next step in the expansion of competitiveness of the U.S. domestic wind energy industry," said Energy Secretary Samuel W. Bodman.[8] Meanwhile, the National Wind Technology Center will also continue operating its blade testing facility.

Other areas of ongoing blade research include sensor-based control systems, materials to withstand harsh marine environments, designs to take full advantage of material use, joints and segmented blades, and lightening protection.[9] The Department of Energy's goal for this research is to stimulate advancements among turbine manufacturers. Commercial deployment of new technologies largely depends on the cost of implementation. For example, blade manufacturers are comfortable with the benefits of carbon fiber and have stepped up their orders for this material. One of the biggest suppliers of carbon fiber to wind turbine manufacturers is St. Louis, Missouri-based Zoltek. According to the company, carbon fiber for turbine blades accounted for 60 to 70 percent of its overall sales in 2007.[10] Increased demand for carbon fiber and other materials for blade manufacturing should encourage new suppliers to enter the business, helping to fill orders and drive down costs. Tight supply situations may cause some turbine manufacturers to explore ways to increase energy output from existing units by implementing dismeter rotors, bend twist coupled blades, and active aerodynamic surfaces.[11]

A General Electric wind turbine blade measuring 121 feet (37 meters) undergoes fatigue testing at the National Wind Technology Center. *Courtesy of the National Renewable Energy Laboratory, Golden, Colorado.*

With larger blade size also comes taller towers. This initiates a new set of logistics criteria for construction and maintenance activities. Turbine manufacturers may need to research the concept of onboard cranes or winches that can remove and lower all components, including gearboxes, generators and blades. Climb assist devices for service technicians will also become increasingly important.

Another important aspect to consider is improved software for overseeing wind farm operations. Each turbine hosts a variety of systems that must work together. Technicians are faced with checking several hundred inputs per turbine, and it only takes one input to be off to shut down an entire turbine. Most of today's turbines operate with almost zero glitches, but as they age, mechanical problems will occur and wind farm operators must be ready to handle them.[12]

Wind farm operators and utilities that increasingly rely on renewable energy sources have turned their attention to developing new power storage technologies, which they hope will significantly diminish inconsistencies between peak winds and electricity demand. While battery and pumped hydro storage have been discussed and tested, there are still significant cost and technical limitations to building these systems for large-scale power requirements. One of the more promising high-capacity wind-based energy storage systems discussed today is underground compressed air. This system works by using off-peak electricity to power compressors which in turn force air into underground storage sites, such as aquifers, spent oil wells, natural caverns and abandoned mines. When the electric demand peaks during the day, the process is reversed. The compressed air is released to the surface, heated by natural gas in combustors, and run through expanders to power an electric generator.[13] The compressed air system thus supplements the power generated by the wind turbines. The Department of Energy, a proponent of compressed air energy storage, noted that "nearly two-thirds of the natural gas in a conventional power plant is consumed by a typical natural gas turbine because the gas is used to drive the machine's compressor. In contrast, a compressed-air storage plant uses low-cost heated compressed air to power the turbines and create off-peak electricity, conserving natural gas."[14]

The National Wind Technology Center conducts a static blade test on a TPI Composites turbine blade measuring 146 feet (44.5 meters). *Courtesy of the National Renewable Energy Laboratory, Golden, Colorado.*

The concept of compressed-air energy storage dates back nearly 30 years. Two natural gas power plants, one in McIntosh, Alabama and another in Huntorf, Germany, pump compressed air into caverns created by salt deposits.[15] Demonstration projects involving compressed air and wind farms have emerged in the U.S. heartland. In 2007, a group of utilities in Iowa, Minnesota and the Dakotas, with support from the Department of Energy, announced plans to integrate a 75-to-150 megawatt wind farm with a compressed air energy storage system outside Des Moines, Iowa. According to the utilities, compressed air at 900 to 1,000 pounds per square inch will be pumped into a sandstone aquifer 3,000 feet below the surface during off-peak hours, such as nights and weekends. The system is designed to generate 268 megawatts of electricity, or enough to power 268,000 average Midwestern homes. The utilities plan to generate electricity through this system at a cost of 6.5 cents per kilowatt hour, and sell it to consumers at peak times for 8 to 10 cents per kilowatt hour. Construction of the compressed air system is scheduled to start in 2009, with completion by 2011.[16] In July 2007, Shell WindEnergy and Luminant, a subsidiary of TXU Corp., said they would explore the use of compressed air storage at their 3,000 megawatt wind farm under construction in the Texas Panhandle.[17] General Compression, an Attleboro, Massachusetts-based company, received about $5 million in start-up funds to develop a wind turbine that directly generates compressed air.[18] The company said its system design, scheduled for prototype in 2009, should eliminate the cost of converting a wind turbine's mechanical power into electricity and then back to mechanical power for the air compressor. Wind energy experts say the drawback with General Compression's design is that the wind turbines will not have the ability to supply electricity directly to the power grid, thus failing to substantiate the cost of the equipment.[19]

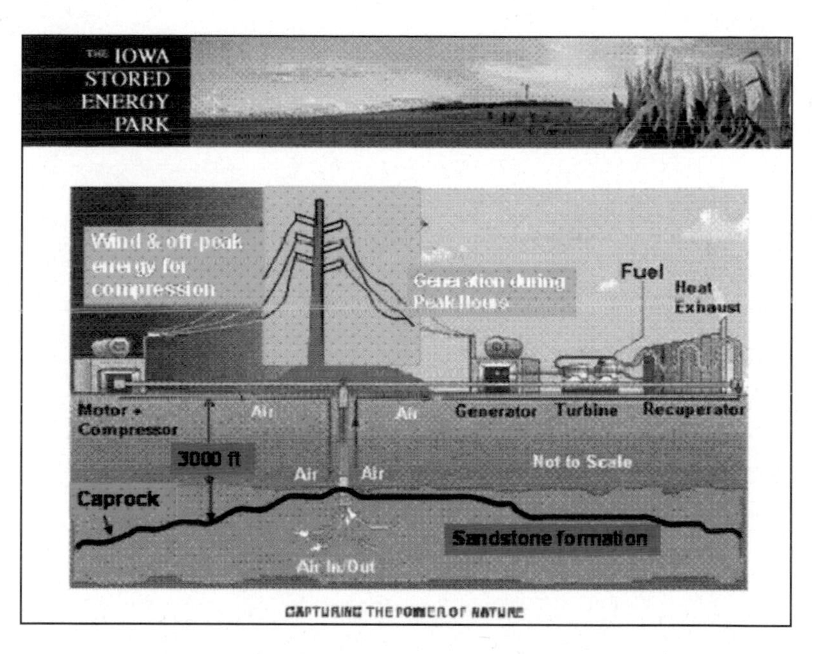


In 2007, a group of utilities in Iowa, Minnesota and Dakotas, with support from the Department of Energy, announced plans to integrate a 75-to-150 megawatt wind farm with a compressed air energy storage system outside Des Moines, Iowa. *Courtesy of The Iowa Stored Energy Park, Ankeny, Iowa, and Wind Utility Consulting, Jefferson, Iowa*.

The first wind energy hydrogen fuel-generating system installed in July 2004 for the Norwegian village Utsira. *Courtesy of Hydro, Oslo, Norway.*

Another promising wind-based energy storage method involves the conversion of wind energy to hydrogen fuel. This system works by using wind turbines during peak operation to send a charge through an electrolyzer to separate hydrogen from oxygen molecules in water. The hydrogen gas is then compressed and stored in tanks for later use. So, when the wind-generated electricity is insufficient to meet peak demand, the hydrogen is released from storage and passed through a fuel cell where an electricity-producing chemical reaction occurs when the hydrogen comes in contact with the oxygen in the air. The first of these systems was turned on in July 2004 by Norwegian power company Hydro in the village of Utsira. The power plant, which includes two 600-kilowatt Enercon wind turbines, immediately began generating sufficient electricity for 10 households.[20] In 2005, ScottishPower in the United Kingdom embarked on a bold five-year plan to introduce wind energy-based hydrogen fuel systems for automobiles. The gas stations to sell the fuel would also be powered by roof-top turbines, according to the plan.[21] In the United States, several wind-to-hydrogen test sites are under development. In 2006, Hydrogenics Corporation of Ontario, Canada, received a contract from the Bismarck, North Dakota-based Basin Electric Power Cooperative to build a wind-based electrolyzer system at North Dakota State University's North Central Research Extension Center at Minot. The cooperative will test the system's ability to provide fuel to vehicles capable of operating on a blend of hydrogen and diesel fuel.[22] In April 2007, the National Renewable Energy Laboratory and Xcel Energy joined forces to roll out the "Wind2H2 project." The hydrogen-generating power plant includes two wind turbines, electrolyzers, four large storage tanks, and a generator to convert the hydrogen to electricity.[23]

Collaboration among the world's prominent renewable energy laboratories will also help take wind turbine technology to new levels of development. The National Renewable Energy Laboratory and Denmark's Risø National Laboratory signed an agreement in June 2007 to cooperate on improving wind energy technologies. The labs will focus on studies of meteorology, aerodynamics, wind turbine structures and materials, control systems, and electrical grid integration. The National Renewable Energy Laboratory in recent years has started focusing on offshore wind turbines and more efficient utility integration. It hopes to share knowledge and experience with the Risø National Laboratory to help advance the technology, taking advantage of many years of research in these areas. "Collaboration among scientists and engineers is of paramount importance. Better solutions often appear when a problem is approached from several different angles," said National Renewable Energy Laboratory Director Dan Arvizu at the time of signing the agreement with Risø National Laboratory.[24]

How far wind turbines will develop in size and scope is yet to be determined. But today's wind farm developers and turbine manufacturers appear open to the possibilities of more change. Wind energy engineer Dayton Griffin described the industry best in a May 2004 article published in *High-Performance Composites* magazine:

> The wind energy industry is presently in its 'teen-age' years – rapidly developing and far from settling into its mature self. Like a teen-ager, the industry has many challenges to face, but almost limitless potential for development. Further advancements in blade design, materials, processes and complementary technologies must play a central role if the wind industry is to realize its full potential.[25]

# Appendices

## Appendix A
## U.S. Wind Farm Map – 2008

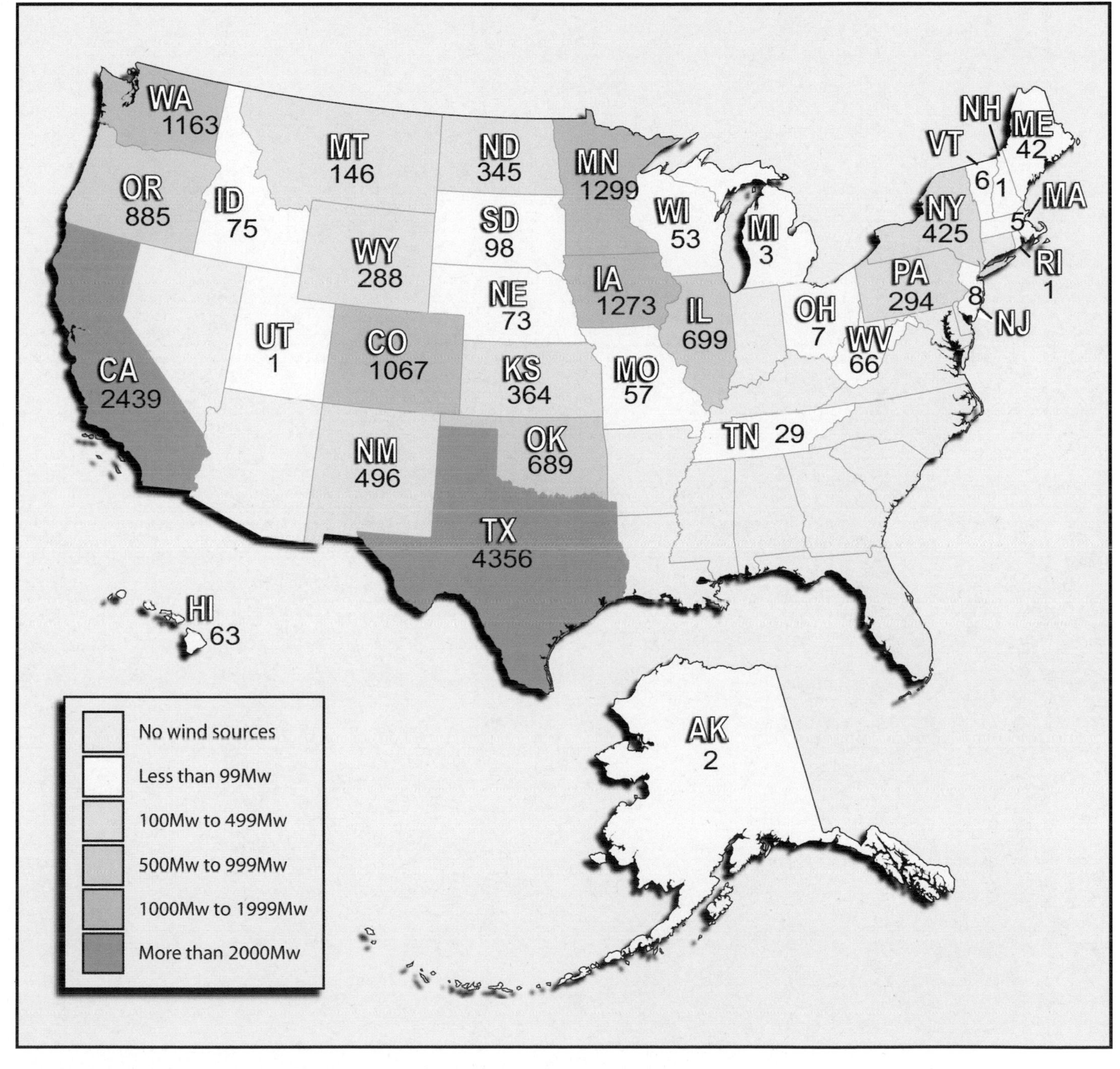


Total installed U.S. wind energy capacity (starting in 1981): 16,818 megawatts as of January, 2008. *Courtesy of the American Wind Energy Association, Washington, D.C.*

# Appendix B

Industry Associations

Global Wind Energy Council
Renewable Energy House
Rue d'Arlon 63-65
1040 Brussels
Belgium
Tel: +32-2-400-1029
Fax: +32-2-546-1944
Web: www.gwec.net

Offshore Wind Energy
Web: www.offshorewindenergy.org

**North America**

American Wind Energy Association
1101 14th Street NW, 12th Floor
Washington, D.C. 20005
Tel: 202-383-2500
Fax: 202-383-2505
Web: www.awea.org

Canadian Wind Energy Association
Suite 810, 170 Lauier Avenue West
Ottawa, Ontario
Canada K1P 5V5
Tel: 613-234-8716
Fax: 613-234-5642
Web: www.canwea.ca

Asociacion Mexicana de Energia Eolica
Jaime Balmes No. 11. L-130F
Col. Los Morales Polanco C.P. 11510 Mexico, D.F.
Tel: +52-55-5395-9559
Web: www.amdee.org

**Europe**

European Wind Energy Association
Rue d'Arlon 63-65
1040 Brussels
Belgium
Tel: +32-2-546-1940
Web: www.ewea.org

EDORA asbl
Rue de la Revolution 7
1000 Brussels
Belgium
Tel: +32-2-217-9682
Fax: +32-2-219-2151
Web: www.edora.be

Danish Wind Industry Association
Vester Voldgade 106
DK-1552 Kobenhavn V
Denmark
Tel: +45-3373-0330
Fax: +45-3373-0333
Web: www.windpower.org

Suomen Tuulivoimayhdistys ry
Raininkaistentie 27
35600 Halli
Finland
Tel: +358-40-771-6114
Web: www.tuulivoimayhdistys.fi

Syndicat des Energies Renouvelables
37, rue La Fayette
75009 Paris
France
Tel: +33-1-48-78-05-60
Fax: +33-1-48-78-09-07
Web: www.enr.fr

Bundesverband Windenergie e.V
Marienstrasse 19-20
10117 Berlin
Germany
Tel: +49-30-284-82-121
Fax: +49-30-284-82-107
Web: www.wind-energie.de

Irish Wind Energy Association
Arigna, Carrick-on-Shannon
Co. Roscommon
Ireland
Tel: +353-71-964-6072
Web: www.iwea.com

Associazione Nazionale Energia del Vento
Via Piemonte, 39
00187 Rome
Italy
Tel: +39-06-4201-4701
Fax: +39-06-4200-4838
Web: www.anev.org

Nederlandse Wind Energie Associatie
Korte Elisabethstraat 6
3511 JG Utrecht
Netherlands
Tel: +31-30-231-6977
Web: www.nwea.nl

Associacao de Energias Renovaveis
Av. Duque de Avila 23
1000-138 Lisbon
Portugal
Tel: +351-213-151-621
Fax: +351-213-151-622
Web: www.apren.pt

Asociacion Empresarial Eolica
Serrano, 143
28006 Madrid
Spain
Tel: +34-917-451-276
Fax: +34-917-451-277
Web: www.aeeolica.org

British Wind Energy Association
Renewable Energy House
1 Aztec Row, Berners Road
London, N1 0PW
United Kingdom
Tel: +44 (0)20-7689-1960
Fax: +44 (0)20-7689-1969
Web: www.bwea.com

**Asia/Africa/Middle East**

Australian Wind Energy Association
Suite 2, Level 6
330 Collins Street
Melbourne VIC 3001
Australia
Tel: +61-3-9670-2033
Fax: +61-3-9602-3055
Web: www.auswind.org

Chinese Renewable Energy Industries Association
No. A2106 Wuhua Plaza
Che Gong Zhuang Street A4
Xi Cheng District
Beijing, 100044
China
Tel: +86-10-6800-2617
Fax: +86-10-6800-2674
Web: www.creia.net

Indian Wind Energy Association
PHD House, 4th Floor
Opp Asian Village
Siri Fort Road
New Delhi 110016
India
Tel: +91-11-2652-3042
Fax: +91-11-2652-3452
Web: www.inwea.org

Indian Wind Turbine Manufacturers Association
417, World Trade Center
Babar Road
New Delhi 110001
India
Tel: +91-44-2441-0333
Web: www.indianwindpower.com

Japanese Wind Energy Association
Kita no Maru Koen 2-1
Chiyoda-ku, Tokyo 102-0091
Japan
Tel: +81-298-58-7275
Web: http://ppd.jsf.or.jp/jwea

Japanese Wind Power Association
Bi-O-Re Akihabara Bldg. 10
Kanda Matsunaga-Cho 18-1
Chiyoda-ku, Tokyo 101-0023
Japan
Tel: +81-3-5297-5577
Fax: +81-3-5297-5578
Web: www.jwpa.jp

New Zealand Wind Energy Association
Level 7, Prime Property Tower
86-90 Lambton Quay
Wellington
New Zealand
Tel: +64-4-499-5046
Web: www.windenergy.org.nz

**South and Central America/Caribbean**

Camara Argentina de Generadores Eolicos
Esmeralda 356 – P9 Of. 29
(C1035ABH) Buenos Aires
Argentina
Tel: +54-11-4328-2551
Web: www.cadge.org.ar

# Appendix C

Small Wind Turbine Equipment Providers

Abundant Renewable Energy
22700 NE Mountain Top Road
Newberg, Oregon 97132
Tel: 503-538-8298
Fax: 503-538-8782
Web: www.abundantre.com

Bergey Windpower Company
2001 Priestley Avenue
Norman, Oklahoma 73069
Tel: 405-364-4212
Fax: 405-364-2078
Web: www.bergey.com

Distributed Energy Systems
182 Mad River Park
Waitsfield, Vermont 05673
Tel: 802-496-2955
Web: www.distributed-energy.com

Entegrity Wind Systems
P.O. Box 832
Charlottetown, PE C1A 7L9
Canada
Tel: 902-368-7171
Web: www.entegritywind.com

Energy Maintenance Service
129 Main Avenue
P.O. Box 158
Gary, South Dakota 57237
Tel: 605-272-5398
Fax: 605-272-5402
Web: www.energyms.com

Lorax Energy
4 Airport Road
Block Island, Rhode Island 02807
Tel: 401-466-2883
Fax: 401-466-2909
Web: www.lorax-energy.com

Solar Wind Works
16713 Greenlee Road
Truckee, California 96161
Tel: 530-582-4503
Fax: 530-582-4603
Web: www.solarwindworks.com

Southwest Windpower
1801 W. Route 66
Flagstaff, Arizona 86001
Tel: 928-779-9463
Fax: 928-779-1485
Web: www.windenergy.com, or www.skystreamenergy.com

Wind Energy Solutions (WES) Canada
2952 Thompson Road
P.O. Box 552
Smithville, Ontario, L0R 2A0
Canada
Tel: 905-957-8971
Fax: 905-957-8789
Web: www.windenergysolutions.ca

Wind Turbine Industries Corporation
16801 Industrial Circle, SE
Prior Lake, Minnesota 55372
Tel: 952-447-6064
Fax: 952-447-6050
Web: www.windturbine.net

# Appendix D

American Windmill and Parts Providers

Aermotor Windmill Co.
P.O. Box 5110
San Angelo, Texas 76902
Tel: 800-854-1656
Fax: 915-651-4948
Web: www.aermotorwindmill.com

Airlift Technologies
11252 Nevada Street
Redlands, California 92373
Tel: 909-446-1780
Web: http://airliftech.com

American Tower Company
5085 Street, Route 39 West
Shelby, Ohio 44875
Tel: 419-347-1185
Fax: 419-347-1654
Web: www.amertower.com

American West Windmill Co.
1701 East 3rd
Amarillo, Texas 79101
Tel: 888-535-4788
Web: www.windmillsupplies.com

American Windmills
5981 Silver Ridge Road
Placerville, California 95667
Tel: 530-644-3008
Web: www.windmills.net

Big Country Windmills
11605 Mairposa Road
Hesperia, California 92345
Tel: 760-244-4828

Big Country Windmills
17386 West Island Road
Maxwell, Nebraska 69151
Tel: 308-582-4451
http://windmills.swnebr.net

Bowjon International Inc.
P.O. Box 610
Bryn Mawr, California 92318
Tel: 909-796-7199
Fax: 909-797-5966

Dakota Windmill & Supply Inc.
28043 SD Hwy 19
Hurley, South Dakota 57036
Tel: 605-238-5110
Web: www.dakotawindmill.com

Dempster Industries
711 South 6th Street

P.O. Box 848
Beatrice, Nebraska 68310
Tel: 402-223-4026
Fax: 402-228-4389
Web: www.dempsterinc.com

Muller Industries
1102 West 21st
Yankton, South Dakota 57078
Tel: 800-316-2727
Fax: 605-665-1925

Outdoor Water Solutions
520 Gage Place
Springdale, Arkansas 72764
Tel: 866-471-1614
Fax: 479-750-9178
Web: www.outdoorwatersolutions.com

Solarwellpumps.com
Rt. 1 Box 52A
Balko, Oklahoma 73931
Tel: 866-483-6851
Web: www.solarwellpumps.com

Texas Windmill
142 S. Riedel Street
Yorktown, Texas 78164
Tel: 361-649-9946

Wewindmills
P.O. Box 266
Pell City, Alabama 35125
Tel: 256-892-1073

WINDTech International
P.O. Box 27
Bedford, New York 10506
Tel: 914-232-2354
Fax: 914-232-2356
Web: www.windmillpower.com

# Appendix E

U.S. Government Sources

Conservation and Production Research Laboratory
Agricultural Research Service
U.S. Department of Agriculture
P.O. Drawer 10
Bushland, Texas 79012-0010
Tel: 806-356-5724
Fax: 806-356-5750
Web: www.cprl.ars.usda.gov

Energy Information Administration
U.S. Department of Energy
1000 Independence Avenue SW
Washington, D.C. 20585
Tel: 202-586-8800
Web: www.eia.doe.gov

National Renewable Energy Laboratory
U.S. Department of Energy
1617 Cole Boulevard
Golden, Colorado 80401-3393
Tel: 303-275-3000
Web: www.nrel.gov

Sandia National Laboratories
P.O. Box 5800
Albuquerque, New Mexico 87185
Tel: 505-284-2001
Web: www.sandia.gov/wind

Wind & Hydropower Technologies Program
Energy Efficiency and Renewable Energy
U.S. Department of Energy
1000 Independence Avenue SW
Washington, D.C. 20585
Tel: 202-586-9220
Web: www.eere.energy.gov

# Appendix F

Publications

*Windpower Monthly*
P.O. Box 100
DK-8420 Knebel
Denmark
Tel: +45-8636-5900
Fax: +45-8636-5626
Web: www.windpower-monthly.com

*WindStats Newsletter*
P.O. Box 100
8420 Knebel
Denmark
Tel: +45-8636-5900
Fax: +45-8636-5626
Web: www.windstats.com

*North American Windpower*
100 Willenbrock Road
Oxford, Connecticut 06478
Tel: 800-325-6745
Fax: 203-262-4680
Web: www.nawindpower.com

*Windtech International*
Siteur Publications
J.C. Kapteynlaan 52
9714 CT Groningen
Netherlands
Tel: +31-50-579-8924
Fax: +31-50-579-8925

*Sun & Wind Energy*
BVA Bielefelder Verlag GmbH & Co. KG
Richard Kaselowsky
Ravensberger Strasse 10f
33602 Bielefeld
Germany
Tel: +49-521-595-514
Fax: +49-521-595-518
Web: www.sunwindenergy.com

*Renewable Energy World*
PennWell International Publications Ltd.
Warlies Park House
Horseshoe Hill, Upshire
Essex EN9 3SR
United Kingdom
Tel: +44-1992-65-6600
Web: www.renewable-energy-world.com

*Wind Today Magazine*
3065 Pershing Ct.
Decatur, Illinois 62526
Tel: 800-728-7511
Fax: 217-877-6647
Web: www.windtoday.net

*Home Power*
P.O. Box 520
Ashland, Oregon 97520
Web: www.homepower.com

# Appendix G

Other Useful Modern Wind Information Sources

Paul Gipe
208 South Green Street, #5
Tehachapi, California 93561-1741
Tel: 661-325-9590
E-mail: pgipe@igc.org
Web: www.wind-works.org

Great Plains Windustry Project
2105 First Avenue South
Minneapolis, Minnesota 55404
Tel: 612-870-3461
Fax: 612-813-5612
Web: www.windustry.org

# Appendix H

Windmill History Associations and Publications

The International Molinological Society
(Publication: *International Molinology*)
TIMS North America
C/O Lisa Steen Riggs
P.O. Box 245
Elk Horn, Iowa 51531-0245
Tel.: 712-764-7472
Fax: 712-764-7475
Web: www.timsmills.info

Society for the Preservation of Old Mills
(Publication: *Old Mill News*)
William L. Denton
Circulation Director
5444 Alpine Ridge
Stevensville, Michigan 49127-1302
Tel: 269-429-0910
E-mail: williamldenton@sbcglobal.net
Web: www.spoom.org

*Windmillers' Gazette*
T. Lindsay Baker, Editor
P.O. Box 507
Rio Vista, Texas 76093
Web: www.windmillersgazette.com

Windmill Study Unit
American Topical Association
(Publication: *Windmill Whispers*)
C/O Charles K. Henderson, North American distributor
623 West Nittany Avenue
State College, Pennsylvania 16801-3967
E-mail: rietta@chilitech.com
Web: www.wsuweb.eu

# Appendix I

American Windmill Museums

American Wind Power Center and Museum
1701 Canyon Lake Drive
Lubbock, Texas 79403
Tel: 806-747-8734
Fax: 806-740-0668
Web: www.windmill.com

Mid-America Windmill Museum
732 S. Allen Chapel Road
Kendalville, Indiana 46755
Tel: 260-897-9918
Web: www.midamericawindmillmuseum.com

# Endnotes

**Chapter One: Origins and History**

[1]Hans E. Wulff, *The Traditional Crafts of Persia: Their Development, Technology, and Influence on Eastern and Western Civilization* (Cambridge, Massachusetts: The M.I.T. Press, 1966), 284-285.

[2]R. Pernoud, *Die Kreuzzuge in Augenseugenberichten* (Dusseldorf, 1961), 363.

[3]Wulff, 285.

[4]Paul Bauters, "The oldest references to windmills in Europe" (paper presented at the fifth annual meeting of The International Molinological Society, 1984).

[5]Frans Brouwers, editor of *Levende Molens*, Ekeren, Belgium, e-mail interview with author, January 3, 2007.

[6]Ibid.

[7]Lynn White Jr., "Medieval Uses of Air," *Scientific American* 243 (August 1970): 97.

[8]Stanley Freese, *Windmills and Millwrighting* (London: Cambridge University Press, 1957), 5.

[9]Rex Wailes, *The English Windmill* (London: Routledge & Kegan Paul Ltd., 1954), 82.

[10]White, "Medieval Uses of Air," 97.

[11]Freese, *Windmills and Millwrighting*, 80.

[12]Ibid.

[13]Brouwers, interview.

[14]Ibid.

[15]Derek Ogden and Anne Burke, "The Windmill at Flowerdew Hundred," *Old Mill News* 6 (January 1978): 4-5.

[16]Philip E. Vierling, *The Fischer Windmill, Elmhurst, Illinois* (Chicago: Illinois Country Outdoor Guides, 1994), 8-9.

[17]Rodney DeLittle, *The Windmills of England* (West Sussex, United Kingdom: Colwood Press Ltd., 1997), 57-59.

[18]Kenneth J. Major, *Watermills and Windmills* (Norwich, United Kingdom: Jarrold and Sons Ltd., 1986), 31.

[19]Roy Gregory, *The Industrial Windmill in Britain* (West Sussex, United Kingdom, Phillimore & Co. Ltd., 2005), 130-135.

[20]Brouwers, interview.

[21]T. Lindsay Baker, "Turbine-Type Windmills of the Great Plains and Midwest," *Agriculture History* 54 (1980): 38.

[22]Ibid, 39.

[23] "A Very High Windmill," *Scientific American* 70 (April 7, 1894): 217.

[24] "A High Windmill," *Scientific American* 70 (May 12, 1894): 292.

[25]Baker, "Turbine-Type Windmills of the Great Plains and Midwest," 45-46.

[26]Douglas R. Hurt, "Irrigation in the West," *Journal of the West* 30 (1991): 63.

[27]T. Lindsay Baker, "Patents as a Key to Understanding Wind Power History in the United States" (paper presented at the eleventh annual meeting of The International Molinological Society, Portugal, September 25-October 2, 2004).

[28]Charles B. Hayward, "Powerful German Windmills," *Scientific American* 92 (March 25,1905): 245-6.

[29]"Homemade Windmills in Nebraska," *Scientific American* 82 (January 13, 1900): 24.

[30]DeLittle, *The Windmills of England*, 12.

[31]Brouwers, interview.

[32]Randall P. Vande Water, "Windmill Celebrates 40th Anniversary!" *The Holland Evening Sentinel,* April 15, 2005.

[33]*Spotcott Windmill* (Cambridge, Maryland: Spocott Windmill Foundation).

[34]Aermotor Windmill Co. price list (San Angelo, Texas, 2005).

**Chapter Two: Pioneers**

[1] "The Sources of Energy in Nature," *Engineering* 32 (1881): 321.

[2] "Mr. Brush's Windmill Dynamo," *Scientific American* 63 (December 20, 1890): 389.

[3] "Brush, Charles Francis," ed. George Wise, in the American National Biography Online, February 2000, http://www.amb.org/articles/13/13-00214.html (accessed January 12, 2007).

[4] "Mr. Brush's Windmill Dynamo," 389.

[5] "City and Ford in Clash over Bush Relic," *The Cleveland Press*, March 14, 1930.

[6]Povl-Otto Nissen, chairman, The Poul la Cour Museum's Friends, "A visit to the Poul la Cour Museum," http://www.povlonis.dk/Plc/visiteng.html (accessed June 15, 2007).

[7] "The Electrical Value of Wind Power," *Scientific American* 93 (November 18, 1905): 394-5.

[8] "Wind-Driven Generators for Farming," *Scientific American* 90 (June 25, 1904): 490.

[9]P.C. Day, "The Winds of the United States and Their Economic Uses," in the *Yearbook of the United States Department of Agriculture* (Washington, U.S. Department of Agriculture, 1911), 337-50.

[10]Putnam A. Bates, "Farm Electric Lighting by Wind Power," *Scientific American* 107 (September 28, 1912): 262.

[11]T. Lindsay Baker, "Wind Electric News: The Papers of Oliver P. Fritchle," *Windmillers' Gazette* 10 (Autumn 1991): 9-10.

[12]Frans Brouwers, "De Meerlann te Gistel," *Levende Molens* 29 (April 2007): 42-46.

**Chapter Three: Small Wind Power**

[1]Mick Sagrillo, "How it all began," *Home Power* 27 (February/March 1992): 15.

[2]Craig Toepfer, Ann Arbor, Michigan, interview with author, March 7, 2007.

[3]Ibid.

[4]Sagrillo, "How it all began," 16.

[5] "Aerodynamic Wind Mills," *Scientific American* 140 (June 1929): 525.

[6]Marcellus L. Jacobs, "Experience with Jacobs Wind Driven Electric Generating Plant – 1931-1957," (personal paper, c. 1957), 1-3.

[7]Ibid., 2.

[8] "The Plowboy Interview: Marcellus Jacobs," *Mother Earth News* 24 (November/December 1973), http://www.motherearthnews.com/UnCategorized/1973-11-01/The-Plowboy-Interview.aspx (accessed January 8, 2007).

[9]Jacobs, "Experience with Jacobs Wind Driven Electric Generating Plant – 1931-1957," 3.

[10]Ibid., 2.

[11]Ibid., 2.

[12] "The Plowboy Interview: Marcellus Jacobs."

[13] "Aerodynamic Wind Mills," 525.

[14]Ron Russell, "Wind-Powered Radio," *Antique Radio Classified* (Web edition) (March 2002), http://www.antiqueradio/Mar02_Russell_Windradio.html (accessed January 16, 2007).

[15]U.S. Department of Energy, Bonneville Power Administration, *BPA Conservation Pilot Project: Wind Power* (February 1981), 1-13.

[16]R. Nolan Clark, agricultural engineer, Conservation and Production Research Laboratory, U.S. Department of Agriculture, Bushland, Texas, interview with author, April 26, 2007.

[17]Toepfer, interview.

[18]Michael L.S. Bergey, president of Bergey Wind Power, Norman, Oklahoma, interview by author, January 2007.

[19]Bergey, interview.

[20]Paul Gipe, "PURPA – A New Law Helps Make Small-Scale Power Production Profitable," *Sierra* 66 (November/December 1981): 54.

[21]Ibid., 55.

[22]Cathy Svejkovsky, ATTRA-National Sustainable Information Service, "Renewable Energy Opportunities on the Farm, 2006," 10-14, http://attra.ncat.org/attra-pub/energyopp.html (accessed July 22, 2007).

[23]Ibid.

[24]Rex A. Ewing, *Power With Nature: Solar and Wind Demystified* (Masonville, Colorado: Pixyjack Press, 2003), 134.

[25]Sara Schaeffer Munoz, "A Novel Way to Reduce Home Energy Bills," *The Wall Street Journal*, August 15, 2006.

[26]American Wind Energy Association, "AWEA Applauds Proposed Incentive for Home, Small Business Wind Systems," February 28, 2007.

[27]Andy Kruse, co-founder of Southwest Windpower, Flagstaff, Arizona, e-mail message to author, January 29, 2007.

[28]American Wind Energy Association, *AWEA Small Wind Turbine Global Market Study 2007*, 2-3.

[29]Toepfer, interview.

[30]Peter Neyens, Notes on Belgian Jos Evens' wind turbine, e-mail to author, May 28, 2007.

[31]Lisa Zyga, "Handheld Windmills Serve As Electric Generators," *PhysOrg.com*, http://www.physorg.com/printnews.php?newsid=90512153 (accessed February 13, 2007).

[32]Claudia Blume, "Hong Kong Inventors Unveil New Micro-Wind Turbines Suitable for City Dwellers," *VOA News*, March 18, 2007, http://www.voanews.org (accessed March 20, 2007).

[33]Chris Gillis and Philip Damas, "Vessel operators ride green wave," *American Shipper* 46 (August 2004): 86.

[34]Mitsui O.S.K. Lines (via Imageline, United Kingdom), e-mail to author, March 12, 2007.

[35]Bergey, interview.

[36]U.S. Department of Energy, National Renewable Energy Laboratory, "DOE's NREL seeks proposals for independent testing for small turbines," July 16, 2007.

**Chapter Four: Scaling Up**

[1] "Plans for a giant windmill," *Science* 61 (March 27, 1925): S10.

[2]Willy Ley, "What Future for Wind Power?" *Science Digest* 26 (August 1954): 84-85.

[3]Palmer Cosslett Putnam, *Power From The Wind* (New York: D. Van Nostrand Company, 1948), 101.

[4]Ibid., 105.

[5]Ibid., 101.

[6]Ibid.

[7]A. Klemin, "Madaras Rotor Power Plant," *Scientific American* 150 (March 1934), 146.

[8]Putnam, 1-2.

[9]Ibid., 105-9.

[10] "Big Windmill," *Fortune* 24 (November 1941): 85.

[11] "Windmill on a Vermont Mountain Top," *Business Week* (May 10, 1941), 22.

[12] "Breezin' through History," *Rutland Herald*, November 4, 2004.

[13]Ibid.

[14]F.A. Annett, "World's Largest Wind-Turbine Plant Nears Completion," *Power* 85 (June 1941): 59.

[15] "Breezin' through History."

[16] "Watts From Wind," *Business Week* (November 24, 1945): 50.

[17]Putnam, 188-89.

[18]Ibid., 175.

[19] "Going with the Wind," *Newsweek* 29 (February 24, 1947): 63.

[20] "Power plant on stilts," *Popular Science* (April 1950): 157.

[21]David Rittenhouse Inglis, *Wind Power and Other Energy Options* (Ann Arbor, Michigan: The University of Michigan Press, 1978), 8.

[22]Matthias Heymann, *Die Geschichte der Windenergienutzung 1890-1990* (Frankfurt, Germany: Campus Verlag, 1995), as reviewed by wind industry expert Paul Gipe, http://www.wind-works.org (accessed September 2, 2007).

[23]Paul N. Vosburgh, *Commercial Applications of Wind Power* (New York: Van Nostrand Reinhold Company, 1983), 29.

[24]Danish Wind Industry Association Website, http://www.windpower.org/en/pictures/juul.htm (accessed March 31, 2007).

[25]Vosburgh, 219.

[26]Tom Kovarik, Charles Pipher, and John Hurst, *Wind Energy* (Northbrook, Illinois: Domus Books, 1979), 16.

[27] "Now There's a Turboprop Windmill," *Popular Science* 167 (September 1955), 177.

[28]British Wind Energy Association, *BWEA 1978-2003 & Beyond*, 19.

[29]Ibid., 19-20.

[30]Ley, "What Future for Wind Power?," 84-85.

[31]Vosburgh, 223.

[32]Kovarik, Pipher, and Hurst, 16-17.

[33]Vosburgh, 223.

[34]Kovarik, Pipher, and Hurst, 17.

[35]George A. Whetstone, "What Can Wind Power Do For Us?," *Power Engineering* 55 (March 1950): 73.

[36]Vosburgh, 224.

[37]NASA Lewis Research Center, Cleveland, Ohio, "Wind Energy Systems: A Non-Pollutive, Non-Depletable Energy," December 1973.

[38] "Back to the windmill to generate power," *Business Week* (May 11, 1974): 140.

[39]David A. Spera, *Bibliography of NASA-Related Publications on Wind Turbine Technology, 1973-1995* (DOE/NASA?5776-3 or NASA CR-195462) (April 1995), 5.

[40]Spera, 9.

[41]Ibid.

[42]Merrill Sheils and Jerry Buckley, "Welcome to 'Monster,'" *Newsweek* 94 (July 23, 1979): 66.

[43]"Noisy Windmill," *Time* 115 (June 2, 1980), 74.

[44]Spera, 12.

[45]Ibid., 13-15.

[46]Ibid., 17-18.

[47]Ibid., 15.

[48]U.S. Department of Energy, Sandia National Laboratories, and American Wind Energy Association, *Vertical Axis Wind Turbines (VAWT): The History of the DOE Program* (1984), 1-3, http://www.sandia.gov/wind/images/VAWThist.html (accessed February 2007).

[49]Ibid., 4-5.

[50]Ibid., 8-9.

[51]William D. Metz, "Wind Energy: Large and Small Systems Competing," *Science* 197 (September 2, 1977): 973.

[52]Paul Gipe, "Mehrkam's windmills," *Popular Science* 214 (April 1979): 82-83.

[53]John Dornberg, "Danish amateurs build the world's biggest windmill," *Popular Science* 214 (January 1979): 81-82.

[54]Ibid., 81.

[55]Allen L. Hammond, "Artificial Tornadoes: A Novel Wind Energy Concept," *Science* 190 (October 17, 1975): 257.

[56]Ben Kocivar, "Tornado turbine reaps power from a whirlwind," *Popular Science* 210 (January 1977): 78.

[57]David Scott, "World's biggest wind machine is a one-armed monster," *Popular Science* 218 (January 1981): 84, 128-29.

**Chapter Five: California's Winds**

[1]Peter Asmus, *Reaping the Wind: How mechanical wizards, visionaries, and profiteers helped shape our energy future* (Washington, D.C.: Island Press, 2001): 61-62.

[2]V. Elaine Smay, "Wind farm's sprout," *Popular Science* 219 (November 1981): 111.

[3]Jim Schefter, "New harvest of energy from wind farms," *Popular Science* 222 (July 1983): 60.

[4]Ibid., 60-61.

[5]Ibid., 61.

[6]Ibid.

[7]Bradford Adams, president of Whitewater Wind Energy, Torrance, California, e-mail interview by author, May 27, 2007.

[8] "Looking in on Altamont's wind farms... by car or on a tour," *Sunset* 174 (April 1985): 78.

[9]Ellen Paris, "The great windmill tax dodge," *Forbes* 133 (March 12, 1984): 40.

[10]Nina Munk, "Mandate power," *Forbes* 154 (August 1, 1994): 41.

[11]Ibid.

[12]Ibid.

[13]Robert W. Righter, *Wind Energy in America: A History* (Norman, Oklahoma: University of Oklahoma Press, 1996), 217.

[14]Bradford Adams, president of Whitewater Wind Energy, Torrance, California, e-mail interview by author, May 21, 2007.

[15]Paul White and Paul Gipe, Comments by

the American Wind Energy Association on repowering California's wind industry for the California Energy Commission's *1994 Biennial Report*, http://www.wind-works.org/articles/Repower.html (accessed April 19, 2007).

[16] "Sitting down with... Fred Noble," *The Desert Sun*, February 18, 2007.

[17]Michael Perrault, "Energy company to cut turbines," *The Desert Sun*, February 21, 2007.

[18]Jim Carlton, "As Demands for Energy Multiply, Wind Farms Stage a Comeback," *The Wall Street Journal*, January 26, 2001.

[19]American Wind Energy Association, "Annual U.S. Wind Power Rankings Track Industry's Rapid Growth," April 11, 2007.

[20]American Wind Energy Association, "U.S. Wind Industry to Install over 3,000 Megawatts of Wind Power in 2007: First Quarter Market Report," May 10, 2007.

[21]Matthew Chew, "Wind power project blows closer," *Tehachapi News*, February 26, 2007.

[22]Perrault, "Energy company to cut turbines."

[23]Mariecar Mendoza, "Commission OKs new windmills," *The Desert Sun*, May 20, 2007.

**Chapter Six: Europe's Leadership**

[1] "Risoe National Laboratory," Danish Wind Industry Association, http://www.windpower.org/en/pictures/eighties.htm (accessed August 11, 2007).

[2] "Modern Wind Turbines," Danish Wind Industry Association, http://www.windpower.org/en/pictures/modern.htm (accessed August 11, 2007).

[3]Global Wind Energy Council, *Global Wind 2006 Report*, 16.

[4]Leila Abboud, "How Denmark Paved Way To Energy Independence," *The Wall Street Journal*, April 16, 2007.

[5] "Power from Wind," *Scientific American* 190 (February 1954): 48.

[6]British Wind Energy Association, *BWEA 1978-2003 & Beyond*, 28-29.

[7]Peter Musgrove, British wind energy expert and advocate, e-mail interview by author, February 17, 2007.

[8]British Wind Energy Association, *BWEA 1978-2003 & Beyond*, 29.

[9]Musgrove, interview.

[10]British Wind Energy Association, *BWEA 1978-2003 & Beyond*, 30.

[11]Musgrove, interview.

[12]Global Wind Energy Council, *Global Wind 2006 Report*, 28.

[13]Ibid, 20.

[14]Ibid, 21.

[15]Ibid, 26.

[16]Crispin Aubrey, "The Spanish Wind Market: Dynamic and Focused," *Wind Directions* (July/August 2005): 16.

[17]Global Wind Energy Council, *Global Wind 2006 Report*, 27.

[18] "The Spanish Wind Market: Global Ambitions," *Wind Directions* (July/August 2005): 30.

[19]Global Wind Energy Council, *Global Wind 2006 Report*, 27.

[20]Ibid, 16.

[21]Ibid, 23.

[22]Crispin Aubrey, "New Team in the European Wind League," *Wind Directions* (July/August 2004): 25.

[23]Mario de Queiroz, "Portugal: Making Up for Lost Time in Renewable Energy," *Inter Press Service News Agency*, June 20, 2007.

[24]Global Wind Energy Council, *Global Wind 2006 Report*, 16.

[25]Christian Kjaer, chief executive of the European Wind Energy Association, "Taking control of our energy future," *EU Power* 2 (2006): 25.

[26]Kimberly Conniff Taber, "Eastern Europe gears up to reap more power from wind," *International Herald Tribune*, June 19, 2007.

[27]Ibid.

[28]Global Wind Energy Council, *Global Wind 2006 Report*, 24.

[29]Ibid, 25.

[30]Taber, "Eastern Europe gears up."

[31]Global Wind Energy Council, *Global Wind 2006 Report*, 25.

[32] "Nike Windfarm in Laakdal," *VincotteKroniek* 4 (February 2007): 27-29.

[33]European Wind Energy Association, "European Market for Wind Turbines Grows 23% in 2006," February 1, 2007.

[34] "EWEA aims for 22% of Europe's electricity by 2030," *Wind Directions* (November/December 2006): 25-26.

[35]Arthouros Zervos, president of the European Wind Energy Association, "Opening remarks" (speech, Seminar Renewables 2020—towards 20%, Lisbon, Portugal, July 11, 2007).

[36] "Wind Power: Main Contributor to 20% EU Target," *Wind Directions* (May/June 2007): 24.

[37] "Interview: Andris Piebalgs, European Energy Commissioner for Energy," *Wind Directions* (May/June 2007): 22.

[38]Ibid.

[39]Ken Silverstein, "Wind Goals No Breeze," *EnergyBiz Insider*, July 27, 2007.

[40]"Wind Power: Main Contributor to 20% EU Target," 24.

[41]European Wind Energy Association, "The Campaign," June 15, 2007, http://www.windday.eu (accessed June 25, 2007).

**Chapter Seven: Building A Wind Farm**

[1]Crystal R. Reid, "No one solution seen to energy issue," *The Bismarck Tribune*, May 22, 2007.

[2]Simon Heaney, "Great Lakes resurgence," *American Shipper* 49 (March 2007): 76.

[3]Port of Vancouver, USA, Washington, "Vestas Signs Exclusive Deal with Port of Vancouver, USA," September 11, 2006.

[4]Port of Longview, Washington, "Port of Longview handles Siemens wind turbines," March 26, 2007.

[5]Port of Olympia, Washington, "Wind energy cargo discharged at Port of Olympia," April 23, 2007.

[6]Mike Allen, "Turbines powering up activity at Port's Tenth Avenue Terminal," *San Diego Business Journal*, May 7, 2007.

[7]Vancouver, USA, September 11, 2006.

[8]Longview, March 26, 2007.

[9]David Ferebee, vice president of sales and marketing, LoneStar Transportation, Fort Worth, Texas, e-mail interview by author, May 1, 2007.

[10]Ibid.

[11] "Green projects," *American Shipper* 47 (August 2005): 22.

[12]Chris Gillis, "Logistics in the wind," *American Shipper* 45 (February 2003): 32.

[13]Ferebee, interview.

[14]Bradford Adams, president of Whitewater Wind Energy, Torrance, California, e-mail interview by author, May 10, 2007.

[15]Crispen Aubrey, "Supply Chain: The race to meet demand," *Wind Directions* (January/February): 27.

[16]American Wind Energy Association, "Wind Power Outlook 2007," http://www.awea.org/pubs/documents/Outlook_2007.pdf. (accessed May 15, 2007).

[17]Ibid.

[18] Ilan Brat, "Crane Migration Hinders Builders," *The Wall Street Journal*, June 18, 2007.

[19]Bradford Adams, president of Whitewater Wind Energy, Torrance, California, e-mail interview by author, May 21, 2007.

[20] Gary Kanaby, Knight & Carver Yacht Center, Inc., National City, California, "Economic Benefits of Scheduled Rotor Maintenance" (paper presented at the American Wind Energy Association Windpower 2006, Pittsburgh, Pennsylvania, June 2006).

[21]Adams, interview, May 21, 2007.

[22] "Oregon community college to start wind energy program," *Associated Press* (posted on *OregonLive.com*), November 19, 2006.

[23] FPL Energy, LLC., Juno Beach, Florida "FPL Energy and Texas College to Train Wind Engineers," February 19, 2007.

[24]Paul Gipe, *Wind Power for Home & Business: Renewable Energy for the 1990s and Beyond*, (White River Junction, Vermont: Chelsea Green Publishing Company, 1993), 338-39.

[25]Paul Gipe, "A Summary of Fatal Accidents in Wind Energy," http://www.wind-works.org/articles/ASummaryofFatalAccidentsinWindEnergy.html. (accessed July 3, 2007).

**Chapter Eight: Offshore Wind**

[1]Danish Wind Industry Association Web site, http://www.windpower.org/en/pictures/offshore.htm (accessed July 23, 2007).

[2]Offshore Wind Energy, http://www.offshorewindenergy.org (accessed September 14,2007).

[3]Walt Musial, Sandy Butterfield, and Bonnie Ram, "Energy from Offshore Wind," (NREL/CP-500-39450, February 2006), 3 (paper presented at the annual Offshore Technology Conference, Houston, Texas, May 1-4, 2006).

[4]Julio Godoy, "Germany Putting More Wind Into Energy," *Inter Press Service News Agency*, February 15, 2007.

[5]Offshore Wind Collaborative Organizing Group, *A Framework for Offshore Wind Energy Development in the United States* (September 2005), 11.

[6]Musial, Butterfield, and Ram, 2-3.

[7]U.S. Department of the Interior, Minerals Management Service, Renewable Energy and Alternative Use Program, *Technology White Paper on Wind Energy Potential on the U.S. Outer Continental Shelf* (May 2006), 2.

[8]Cape Wind, "Project at a glance," http://www.capewind.org/printarticle24.htm (accessed on November 2, 2005).

[9]David Tyler, "World's largest offshore wind farm proposed near Cape Cod," *Professional Mariner* 69 (December/January 2003): 9.

[10]The Commonwealth of Massachusetts, Executive Office of Environmental Affairs, "Secretary Bowles Signs MEPA Certificate Finding Cape Wind Environmental Review 'Adequate,'" March 30, 2007.

[11] "Final Decision On Project To Be Made In 2008," *TheBostonChannel.com*, April 6, 2007.

[12]Stephanie Ebbert, "Cape Wind moves on to federal review," *The Boston Globe*, March 31, 2007.

[13]Frank Eltman, "N.Y. utility scrapping ocean wind park," *Associated Press*, August 24, 2007.

[14]Rachel Swick, "Bluewater Wind of New York unveils wind power proposal," *Cape Gazette*, January 26, 2007.

[15]Texas General Land Office, "Texas lands historic offshore wind project," October 24, 2005.

[16]Texas General Land Office, "Patterson signs lease for biggest offshore wind farm in U.S. history," May 11, 2006.

[17]John Porretto, "Offshore wind farm plans are called off," *Associated Press* (posted on *Star-Telegram.com*), June 13, 2007.

[18]Robert H. Owen Jr., president of Superior Safety and Environmental Services Inc., Middleton, Wisconsin, *Final Report to Wisconsin Focus on Energy on Lake Michigan Offshore Wind Resource Assessment* (funded by the Wisconsin Focus on Energy Program) (July 30, 2004), 61.

[19]Chris Gillis, "Offshore logistics specialist," *American Shipper* 49 (May 2007): 22.

[20]Ibid., 24.

[21] "World's first wind and gas offshore energy project given green light," *PRNewswire-GNN*, February 8, 2007.

[22]Eddie O'Connor, chief executive officer and founder of Airtricity, Dublin, Ireland, "The European Offshore Supergrid," *Windtech International* 3 (January/February 2007): 8.

[23]Ibid., 9.

[24]Nancy Stauffer, "Deep-sea oil rigs inspire giant wind turbines," *MIT Tech Talk* 51 (September 13, 2006): 4.

[25]Ibid.

[26]Ibid.

**Chapter Nine: North America's Wind Rush**

[1] "Green logistics," *American Shipper* 47

(August 2005): 22.

[2] "Farmers who inherit the wind being paid for it," *New York Times News Service* (Published in *The Sun* of Baltimore, Maryland), November 26, 2000.

[3] "Green logistics," August 2005, 22.

[4]GE Power Systems, "GE Power Systems signs agreement to acquire Enron's wind business," February 20, 2002.

[5]American Wind Energy Association, "Boom: 2003 close to best ever year for new wind installations; Bust: Expiration of key incentive lowers hopes for 2004," January 22, 2004.

[6]American Wind Energy Association, "First quarter market report: Wind industry trade group sees little to no growth in 2004, following near-record expansion in 2003," May 12, 2004.

[7]American Wind Energy Association, "Second quarter market report: American wind industry needs consistent business environment to bring wind power's promise to the country," August 10, 2004.

[8]American Wind Energy Association, "Energy bill extends wind power incentive through 2007," July 29, 2005.

[9]Rebecca Smith, "States Lead Renewable-Energy Push," *The Wall Street Journal*, September 22, 2004.

[10]Mike Sloan, "The Texas RPS: Gusher, dry hole – or both?," Earthscan http://www.earthscan.co.uk, February 21, 2005 (accessed May 17, 2005).

[11]American Wind Energy Association, "Renewable portfolio standard adopted by New York state," September 24, 2004.

[12]Ibid.

[13]American Wind Energy Association to U.S. Congressional Leadership, letter, "Renewable Energy: The Time Has Come," May 24, 2007.

[14]The Rosebud Sioux, Rosebud, South Dakota, "First Native American-owned utility-scale wind turbine to be dedicated," April 22, 2003.

[15]David Melmer, "Tribal energy organization wins worldwide recognition," *Indian Country Today*, July 9, 2007.

[16]Jill K. Cliburn, "Promising New Crop," *Rural Electric* 62 (November 2004): 40-43.

[17] "Trimont Area Wind Farm Celebrates Dedication; Nation's Largest Landowner-Developed Wind Farm Generates Enough Electricity to Serve the Annual Energy Needs of about 29,000 Homes," *Business Wire*, July 7, 2007.

[18]American Wind Energy Association, "Annual U.S. wind power rankings track industry's rapid growth," April 11, 2007.

[19]American Wind Energy Association, August 10, 2004.

[20]American Wind Energy Association, "Wind power development can create thousands of manufacturing jobs in states hit hardest by recent job losses, study finds," October 13, 2004.

[21]Matt Phinney, "Wind energy may boost small towns," *San Angelo Standard-Times*, August 10, 2006.

[22]UPC Wind, Newton, Massachusetts, "Steel Winds project achieves full commercial operations," June 5, 2007.

[23]Maki Becker, "Windmill power spins into reality," *The Buffalo News*, June 7, 2007.

[24]Maki Becker, "Towers of power along the lake," *The Buffalo News*, August 20, 2006.

[25]Kristyn Ecochard, "Analysis: Ridge-top wind generates debate," *UPI*, March 27, 2007.

[26]American Wind Energy Association, "U.S. wind energy industry applauds federal energy regulatory commission ruling on new transmission policy," April 19, 2007.

[27] "Largest wind farm in North Dakota announced," *PR-inside.com*, March 29, 2007.

[28]American Wind Energy Association, "Installed U. S. wind power capacity surged 45% in 2007: American Wind Energy Association Market Report," January 17, 2008.

[29]American Wind Energy Association, April 11, 2007.

[30]FPL Energy, LLC, "FPL Energy announces

expanded growth plan for wind business," July 30, 2007.

[31]American Wind Energy Association, "U.S. wind industry to install over 3,000 megawatts of wind power in 2007: First quarter market report," May 10, 2007.

[32]Shell WindEnergy, "Luminant and Shell join forces to develop a Texas-sized wind farm," July 27, 2007.

[33]Betsy Blaney, "Pickens wants to build world's biggest wind farm," *Associated Press*, June 13, 2007, http://www.myfoxfw.com (accessed June 15, 2007).

[34]Randall Swisher, president of the American Wind Energy Association, telephone interview by author, August 24, 2007.

[35]Global Wind Energy Council, *Global Wind 2006 Report*, 32.

[36]Canadian Wind Energy Association, "2006 a record breaking year for the global wind energy industry with Canada now ranking 12th in the world for total installed wind capacity," February 8, 2007.

[37]Marianne White, "Quebec moves to forefront of Canada's rapidly growing wind industry," *CanWest News Service* (Posted on http://www.canada.com), July 27, 2007.

[38]Global Wind Energy Association, *Global Wind 2006 Report*, 33.

[39]Natural Resources Canada, Calgary, Alberta, "Canada's Government announces $16.5 million to Kettles Hill wind energy project," July 5, 2007.

[40]Ibid.

[41]Global Wind Energy Council, *Global Wind 2006 Report*, 36-37.

**Chapter Ten: Global Spread**

[1]Global Wind Energy Council, *Global Wind 2006 Report*, 13.

[2]Ibid., 38.

[3]Ibid., 38-39.

[4]Ibid., 39.

[5]Ibid.

[6] "Tanti capitalizes on global warming concerns," *PTI* (posted in *The Economic Times Online*), May 27, 2007.

[7]Ibid.

[8] Global Wind Energy Council, *Global Wind 2006 Report*, 39.

[9]Ibid., 40-41.

[10]U.S. Department of Energy, National Renewable Energy Laboratory, *Renewable Energy in China* (NREL/FS-710-35789) (April 2004), http://www.nrel.gov/international.

[11]Charlie Dou, China manager for Bergey Windpower, Beijing, e-mail interview by author, March 1, 2007.

[12]Global Wind Energy Council, *Global Wind 2006 Report*, 40.

[13]Howard W. French, "In search of a new energy source, China rides the wind," *The New York Times*, July 26, 2005.

[14]Global Wind Energy Council, *Global Wind 2006 Report*, 40.

[15]Ibid., 40-41.

[16] "Wind energy cuts CO2 emissions in Taiwan by 250,000 tons a year," *The China Post* (Internet Edition), April 23, 2007, http://www.chinapost.com.tw/news/print/107807.htm (accessed April 24, 2007).

[17]Ibid.

[18]Ibid.

[19] "Wind Power on the Increase in Japan," *JETRO-Japan Economic Report*, April-May 2006.

[20]Global Wind Energy Council, *Global Wind 2006 Report*, 42.

[21] "Wind Power Takes Off," *Web Japan*, March 29, 2006, http://web-japan.org/trends/business/bus060329.html (accessed April 23, 2007).

[22]Mitsubishi Heavy Industries, Ltd., Tokyo, Japan, "MHI to triple wind turbine production capacity to 1,200 MW/year," April 10, 2007.

[23] "Wind Power on the rise in South Korea," *Energy Daily News* (Seoul, South Korea), (posted by *RenewableEnergyAccess.com*), January 15, 2007.

[24]Global Wind Energy Council, *Global Wind 2006 Report*, 44-45.

[25]International Energy Agency, *2005 Annual Report*, "Chapter 20 Korea," 175-76.

[26] "Wind power on the rise in South Korea," *Energy Daily News*, January 15, 2007.

[27] "North Korea focusing on developing wind energy," *Yonhap News*, November 23, 2006.

[28] "Building a political bridge with wind," *Windpower Monthly* (May 1999), http://www.nautilus.org/archives/dprkrenew/wpowerarticle.html (accessed May 7, 2007).

[29]The Nautilus Institute, Berkley, California, "Nautilus Institute completes first American wind power village project in North Korea," October 6, 1998.

[30]Ibid.

[31]The Nautilus Institute, "Nautilus concludes third renewable energy mission to the Democratic People's Republic of Korea," October 3, 2000.

[32]Peter Hayes, executive director of The Nautilus Institute, Berkley, California, e-mail interview by author, April 29, 2007.

[33]Ibid.

[34]*Yonhap News*, November 23, 2006.

[35]Helen Ubels, "Australia pulls plug on support for its wind-power program," (*Dow Jones Newswires*) *The Wall Street Journal*, June 15, 2005.

[36]Global Wind Energy Council, *Global Wind 2006 Report*, 7.

[37]Ibid., 46.

[38]Ibid., 7.

[39]Ibid., 50.

[40]Ibid., 51.

[41]Ibid., 48.

[42]Ibid., 49.

[43]Ibid., 52-53.

[44] "Dutch firm to roll out US$160 million wind energy project," *Africa News*, April 25, 2007.

[45]Kui Kinyanjui, "Wind power project to boost energy supply," *Business Daily Africa*, June 15, 2007.

[46]Monte Reel, "Argentine town hopes to transform wind into windfall," *The Washington Post*, May 15, 2006.

[47]Global Wind Energy Council, *Global Wind 2006 Report*, 34-35.

[48]Matt Malinowski, "Wind energy catches on in Chile," *The Santiago Times*, August 20, 2007.

[49] "Cuba testing wind energy," *Prensa Latina*, June 11, 2007 (posted in *ahora.cu*), http://www.ahora.cu (accessed June 12, 2007).

**Chapter Eleven: Fitting In**

[1]Martin Beckford, "Wind turbines 'are ruining our quality of life,'" *The Telegraph*, April 16, 2007.

[2]American Wind Energy Association, "New national coalition formed to promote clean, renewable, domestic wind energy development," July 6, 2005.

[3]Alice Ross, "Debunking the myths: Countering the common objections leveled at wind turbines," *Refocus* 7 (May 2006): 41.

[4] "Windmills in my backyard," *The Copenhagen Post*, April 25, 2007.

[5]The National Academies, "Use of wind energy in U.S. growing, but planning and guidelines are lacking," May 3, 2007.

[6] "Blades," *Audubon* 92 (May 1990): 21.

[7]California Energy Commission, Public Interest Energy Research Program: Final Report, *Developing Methods to Reduce Bird Mortality in the Altamont Pass Wind Resource Area* (April 2004), http://www.energy.ca.gov/pier/final_project_reports/500-04-052.html (accessed August 18, 2004).

[8]Ibid.

[9]Ibid.

[10]Justin Blum, "Researchers alarmed by bat deaths from wind turbines," *The Washington Post*, January 1, 2005.

[11]U.S. Government Accountability Office, *Wind Power: Impacts on Wildlife and Government Responsibilities for Regulating Develop-*

*ment and Protecting Wildlife* (GAO-05-906) (September 2005), 4.

[12]House Committee on Natural Resources, Subcommittee on Fisheries, Wildlife and Oceans, "Gone with the Wind: Impacts of Wind Turbines on Birds and Bats," May 1, 2007 (Testimony by H. Dale Hale, director of the U.S. Fish and Wildlife Service).

[13]U.S. Department of the Interior, Office of the Secretary, "Establishment of Wind Turbine Guidelines Advisory Committee," *Federal Register* 72 (March 13, 2007): 11373-11374.

[14]Hale testimony, House Committee on Natural Resources, May 1, 2007.

[15]House Committee on Natural Resources, Subcommittee on Fisheries, Wildlife and Ocean, "Gone with the Wind: Impacts of Wind Turbines on Birds and Bats," May 1, 2007 (Testimony by Mike Daulton, director of conservation policy, National Audubon Society, and Donald Michael Fry, director of pesticides and birds program, American Bird Conservancy).

[16]House Committee on Natural Resources, Subcommittee on Fisheries, Wildlife and Ocean, "Gone with the Wind: Impacts of Wind Turbines on Birds and Bats," May 1, 2007 (Opening statement by Representative Nick J. Rahall II of West Virginia, chairman of the House Committee on Natural Resources).

[17]The National Academies, May 3, 2007.

[18] "Wind energy suspension for the birds," *The Anatolia News Agency* (posted in the *Turkish Daily News*), February 20, 2007.

[19]Fry testimony, House Committee on Natural Resources, May 1, 2007.

[20]Don Hopey, "Aviary tracking raptors to find safe sites for wind turbines," *Pittsburgh Post-Gazette*, January 14, 2007.

[21]Peter Asmus, "Who owns the wind?: Wind power pits clean energy against endangered birds," *E; the Environmental Magazine* 4 (May/June 1993): 18-20.

[22]Mike Tidwell, director of the Chesapeake Climate Action Network, letter to the editor, *The Washington Post*, February 6, 2005.

[23]American Wind Energy Association, July 6, 2005.

**Chapter Twelve: Growing Up**

[1]Bradford Adams, president of Whitewater Wind Energy, e-mail interview by author, July 18, 2007.

[2]Thomas Ashwill and Daniel Laird, Sandia National Laboratories, Albuquerque, New Mexico, "Concepts to Facilitate Very Large Blades," (paper presented at the 45th AIAA Aerospace Sciences Meeting and Exhibit, Reno, Nevada, January 8-11, 2007).

[3] "Wind Turbine Blade," Sandia National Laboratories, http://sandia.gov/winf/Blades.html (accessed August 24, 2007).

[4]Tom Ashwill, Sandia National Laboratories, Albuquerque, New Mexico, "Blade Technology Innovations for Utility-Scale Turbines," (paper presented at the American Wind Energy Association Windpower 2006, Pittsburgh, Pennsylvania, June 2006).

[5]Ibid.

[6]U.S. Department of Energy, "Department of Energy to Invest up to $4 million for Wind Turbine Blade Testing Facilities," June 25, 2007.

[7]National Renewable Energy Laboratory, "Large Wind Turbine Blade Test Facilities to be in Mass., Texas," June 25, 2007.

[8]U.S. Department of Energy, June 25, 2007.

[9]Ashwill, "Blade Technology Innovations for Utility-Scale Turbines."

[10]Marilyn Alva, "For Big Carbon-Fiber Makers, The Answer is Blowing In The Wind," *Investor's Business Daily*, June 7, 2007.

[11]Adams, interview.

[12]Ibid.

[13] "Compressed Air Energy Storage," U.S. Department of Energy, http://www.eere.energy.gov/de/compressedair.html (accessed August 3, 2007).

[14]Ibid.

[15]Ibid.

[16]Mike Meyers, "Saving up the wind's energy," *Star Tribune*, April 2, 2007.

[17]Shell WindEnergy, "Luminant and Shell Join Forces to Develop a Texas-Sized Wind Farm," July 27, 2007.

[18]General Compression, "Pioneer of Dispatchable Wind Technology raises over $5,000,000," March 2007.

[19] "Trapped Wind," *The Economist* 384 (July 26, 2007): 82.

[20]Hydro, "Utsira wind power and hydrogen plant inauguration," July 1, 2004.

[21]Paul Lamarra and Mark Macaskill, "Spare wind turbine power to fuel hydrogen cars," *The Sunday Times*, August 21, 2005.

[22]James MacPhearson, "Wind-to-hydrogen project to be dedicated," *Associated Press*, July 18, 2007.

[23]Julia Thomas, "'Wind to Hydrogen' Facility Offers New Template for Future Energy Production," National Renewable Energy Laboratory, April 2007, http://www.nrel.gov/features/04-07_xcel_wind_hydro.html (accessed August 30, 2007).

[24]National Renewable Energy Laboratory, "U.S., Danish laboratories to cooperate on wind energy research," June 27, 2007.

[25]Dayton Griffin, "Growing opportunities and challenges in wind turbine blade manufacturing," *High-Performance Composites* (May 204), http://www.compositesworld.com/hpc/issues/2004/May/450 (accessed August 24, 2007).

# Glossary

*Courtesy of Horizon Wind Energy, Houston, Texas*

**Air Pollution** - Air with contaminants in it that prevent the air from dispersing as it normally would, and interfere with biological processes.

**Alternative Energy** - A popular term for "non-conventional" energy like renewables.

**Asynchronous Generator** - A type of electric generator that produces alternating current (AC) electricity to match an existing power source.

**Battery** - An energy storage device made up of one or more electrolyte cells. An electrolyte is a nonmetallic conductor that carries current.

**Breeze** - Wind classified as light, gentle, moderate, fresh, or strong.

**Carbon Dioxide (CO2)** - A colorless, odorless noncombustible gas present in the atmosphere. It is formed by the combustion of carbon and carbon compounds (such as fossil fuels and biomass), by respiration, which is a slow combustion in animals and plants, and by the gradual oxidation of organic matter in the soil. It is a greenhouse gas that contributes to global climate change.

**Carbon Monoxide (CO)** - A colorless, odorless but poisonous combustible gas. Carbon monoxide is produced in the incomplete combustion of carbon and carbon compounds, for example, fossil fuels like coal and petroleum.

**Central Power Plant** - A large power plant that generates power for distribution to multiple customers.

**Chemical Energy** - The energy liberated in a chemical reaction, as in the combustion of fuels.

**Circuit** - A device, or system of devices, that allows electrical current to flow through and voltage to occur across positive and negative terminals.

**Circuit Breaker** - A device used to interrupt or break an electrical circuit when an overload condition exists. Circuit breakers are used to protect electrical equipment from potential damage.

**Climate** - The prevailing or average weather conditions of a geographic region.

**Conductor** - The material through which electricity is transmitted, such as an electrical wire.

**Conduit** - A tubular material used to encase and protect electrical conductors.

**Constant-Speed Wind Turbines** - Wind turbines that operate at a constant rpm (rotor revolutions per minute). They are designed for optimal energy capture at a specific rotor diameter and at a particular wind speed.

**Conventional Fuel** - The fossil fuels: coal, oil, and natural gas.

**Converter** - A device for transforming electricity to a desired quality and quantity.

**Cycle** - In alternating current electricity, the current flows in one direction from zero to a maximum voltage, then goes back down to zero, then to a maximum voltage in the opposite direction. This comprises one cycle. The number of complete cycles per second determines the current frequency. In the United States the standard for alternating current is 60 cycles.

**Cyclone** - Air spinning inward toward centers of low air pressure. Cyclones spin counterclockwise in the Northern Hemisphere and clockwise in the Southern Hemisphere.

**Deregulation** - The process of changing policies and laws of regulation in order to increase competition among suppliers of commodities and services. The Energy Policy Act initiated deregulation of the electric power industry in 1992.

**Direct Current** - A type of electricity transmission and distribution by which electricity flows in one direction through the conductor. Usually the electricity is a relatively low voltage and high current. Direct current is abbreviated as DC.

**Distribution** - The process of distributing electricity. Distribution usually refers to the portion of power lines between a utility's power pole and transformer and a customer's point of connection.

**Doldrums** - A narrow, virtually windless zone near the Equator, created as heated air rises upward, leaving the ocean's surface calm and glassy.

**Downburst** - A severe localized downdraft from a thunderstorm. Also called a microburst.

**Downwind Wind Turbine** - A horizontal axis wind turbine in which the rotor is downwind of the tower.

**Electricity** - The energy of moving electrons, the current of which is used as a source of power.

**Electricity Generation** - The process of producing electricity by transforming other forms or sources of energy into electrical energy. Electricity is measured in kilowatt hours (kWh).

**Emission** - A substance or pollutant emitted as a result of a process.

**Energy** - The capacity for work. Energy can be converted into different forms, but the total amount of energy remains the same.

**Energy Storage** - The process of storing or converting energy from one form to another for later use. An example of a storage device is a battery.

**Environment** - All the natural and living things around us. The earth, air, weather, plants, and animals all make up our environment.

**Frequency** - The number of cycles through which an alternating current passes per second, measured in hertz.

**Fuel** - Any material that can be consumed to make energy.

**Gearbox** - A protective casing for a system of gears.

**Generator** - A device for converting mechanical energy to electrical energy.

**Gigawatt (GW)** - A unit of power equal to 1 million kilowatts.

**Global Warming** - A term used to describe the increase in average global temperatures due to the greenhouse effect.

**Green Credit** - Green credits are a new way to purchase renewable electric generation that divides the generation into two separate products: the commodity energy and the renewable attributes. The green credit represents the renewable attributes of a single megawatt of renewable energy. Also known as green tags, renewable energy credits, or renewable energy certificates.

**Green Power** - A popular term for energy produced from renewable energy resources.

**Greenfield** - A site on which a power plant has not previously existed.

**Greenhouse Effect** - The heating effect resulting from long wave radiation from the sun being trapped by greenhouse gases that have been produced from natural and human sources.

**Greenhouse Gases** - Gases such as water vapor, carbon dioxide, methane, and low-level ozone that are transparent to solar radiation, but opaque to long wave radiation. These gases contribute to the greenhouse effect.

**Grid** (also "**Power Grid**" and "**Utility Grid**") - A common term referring to an electricity transmission and distribution system.

**Gust** - A sudden brief increase in the speed of the wind.

**Hertz (Hz)** - A measure of the number of cycles or wavelengths of electrical energy per second. The United States electricity supply has a standard frequency of 60 hertz.

**Horizontal-Axis Wind Turbines** - Turbines on which the axis of the rotor's rotation is parallel to the wind stream and the ground.

**Jet Stream** - A meandering and relatively narrow belt of strong winds embedded in the normal wind flow, generally found at high altitudes.

**Joule (J)** - A metric unit of energy or work. One joule per second equals 1 watt.

**Kilowatt (kW)** - A standard unit of electrical power equal to 1,000 watts.

**Kilowatt-Hour (kWh)** - A unit or measure of electricity supply or consumption of 1,000 watts over the period of one hour.

**Kinetic Energy** - Energy available as a result of motion. (Kinetic energy is equal to one half the mass of the body in motion multiplied by the square of its speed.)

**Knot** - One nautical mile per hour (1.15 MPH).

**Landman** - An in-house or independent land management consultant who negotiates terms of land leases with land owners.

**Leading Edge** - The surface part of a wind turbine blade that first comes into contact with the wind.

**Lift** - The force that pulls a wind turbine blade.

**Mean Power Output** (of a Wind Turbine) - The average power output of a wind energy conversion system at any given mean wind speed.

**Mean Wind Speed** - The average wind speed over a specified time period and height above the ground.

**Mechanical Energy** - The energy possessed by an object due to its motion (kinetic energy) or its potential energy.

**Median Wind Speed** - The wind speed with 50% probability of occurring.

**Megawatt (MW)** - The standard measure of electric power plant generating capacity. One megawatt is equal to one thousand kilowatts or 1 million watts.

**Megawatt-hour (MWh)** - 1,000 kilowatt-hours or 1 million watt-hours.

**Met Tower** - Meteorological towers erected to verify the wind resource found within a certain area of land.

**Nitrogen Oxides (NOx)** - The products of all combustion processes formed by the combination of nitrogen and oxygen. Nitrogen oxides and sulfur dioxide are the two primary causes of acid rain.

**Non-Renewable Fuels** - Fuels that cannot be easily renewed or reproduced, such as oil, natural gas, and coal.

**Peak Wind Speed** - The maximum instantaneous wind speed that occurs within a specific period of time.

**Power** - Energy that is capable or available for doing work.

**Power Grid** (also **"Utility Grid"**) - A common term referring to an electricity transmission and distribution system.

**Power Quality** - Stability of frequency and voltage and lack of electrical noise on the power grid.

**Prevailing Wind Direction** - The direction from which the wind predominantly blows as a result of the seasons, high and low pressure zones, the tilt of the earth on its axis, and the rotation of the earth.

**Renewable Energy** - Energy derived from resources that are regenerative or that cannot be depleted. Types of renewable energy resources include wind, solar, biomass, geothermal, and moving water.

**Restructuring** - The process of changing the structure of the electric power industry from one of a guaranteed monopoly that is regulated to one of open competition between power suppliers.

**Solar Energy** - Electromagnetic energy transmitted from the sun (solar radiation).

**Step-Up Gearbox** - A step-up gearbox increases turbine electricity production in stages by increasing the number of generator revolutions produced by the rotor revolutions.

**Sulfur Dioxide (SO2)** - A colorless gas released as a by-product of combusted fossil fuels containing sulfur. The two primary sources of acid rain are sulfur dioxide and nitrogen oxides.

**Sustainable Energy** - Energy that takes into account present needs while not compromising the availability of energy or a healthy environment in the future.

**Trade Wind** - The consistent system of prevailing winds occupying most of the tropics. They constitute the major component of the general circulation of the atmosphere. Trade winds blow northeasterly in the Northern Hemisphere and southeasterly in the Southern Hemisphere. The trades, as they are sometimes called, are the most persistent wind system on earth.

**Turbine** - Also see "Wind Turbine." A term used for a wind energy conversion device that produces electricity.

**Turbulence** - A swirling motion of the atmosphere that interrupts the flow of wind.

**Utility Grid** - Also see "Power Grid." A common term referring to an electricity transmission and distribution system.

**Variable-Speed Wind Turbines** - Turbines in which

the rotor speed increases and decreases with changing wind speeds. Sophisticated power control systems are required on variable speed tubines to insure that their power maintains a constant frequency compatible with the grid.

**Volt** - A unit of electrical force.

**Voltage** - The amount of electromotive force, measured in volts, that exists between two points.

**Watt (W)** - The rate of energy transfer (from an outlet to an appliance, for example). Wattage is calculated by multiplying voltage by current.

**Watt-Hour (Wh)** - A unit of electricity consumption of one watt over the period of one hour.

**Wind** - Moving air. The wind's movement is caused by the sun's heat, the earth, and the oceans, forcing air to rise and fall in cycles.

**Wind Energy** - (Also see "Wind Power") Power generated by converting the mechanical energy of the wind into electrical energy through the use of a wind generator.

**Wind Energy Conversion System (WECS)** - An apparatus for converting wind energy to mechanical energy, making it available for powering machinery and operating electrical generators.

**Wind Farm** - A piece of land on which wind turbines are sited for the purpose of electricity generation.

**Wind Generator** - A wind energy conversion system designed to produce electricity.

**Wind Power** - ( Also see "Wind Energy") Power generated by converting the mechanical energy of the wind into electrical energy through the use of a wind generator.

**Wind Power Plant** - A group of wind turbines interconnected to a common utility system.

**Wind Resource Assessment** - The process of characterizing the wind resource and its energy potential for a specific site or geographical area.

**Wind Rose** - A diagram that indicates the average percentage of time that the wind blows from different directions, on a monthly or annual basis.

**Wind Speed** - The rate of flow of wind when it blows undisturbed by obstacles.

**Wind Speed Frequency Curve** - A curve that indicates the number of hours per year that specific wind speeds occur.

**Wind Speed Profile** - A profile of how the wind speed changes at different heights above the surface of the ground or water.

**Wind Turbine** - A term used for a wind energy conversion device that produces electricity.

**Wind Turbine Rated Capacity** - The amount of power a wind turbine can produce at its rated wind speed.

**Wind Velocity** - The wind speed and direction in an undisturbed flow.

**Windmill** - A wind energy conversion system that is used to grind grain. However, the word windmill is commonly used to refer to all types of wind energy conversion systems.

**Windpower Profile** - The change in the power available in the wind due to changes in the wind speed or velocity.

# Index